Humanism Revisited

HUMANISM REVISITED

An Anthropological Perspective

Rik Pinxten

berghahn
NEW YORK · OXFORD
www.berghahnbooks.com

First published in 2024 by
Berghahn Books
www.berghahnbooks.com

© 2024 Rik Pinxten

All rights reserved. Except for the quotation of short passages
for the purposes of criticism and review, no part of this book
may be reproduced in any form or by any means, electronic or
mechanical, including photocopying, recording, or any information
storage and retrieval system now known or to be invented,
without written permission of the publisher.

Library of Congress Cataloging-in-Publication Data

A C.I.P. catalogue record for this book is available from the Library of Congress.
Library of Congress Cataloging in Publication Control Number:
2023051020

British Library Cataloguing in Publication Data

A catalogue record for this book is available from the British Library

ISBN 978-1-80539-473-0 hardback
ISBN 978-1-80539-474-7 epub
ISBN 978-1-80539-475-4 web pdf
https://doi.org/10.3167/9781805394730

Dedicated to Ellen, my soulmate and loved one. She convinced me that each human being should try and make a difference.

CONTENTS

Foreword *Laura Nader*		viii
Preface		x
Note on Text. Warning		xx
Introduction. The Zebra and the Dolphin in Us		1
Part I. The State of Things		
Chapter 1.	Humanism Today: What Are We Doing, and What Is the Context?	11
Chapter 2.	Worldwide Interdependence in the Twenty-First Century	28
Part II. A Plea for NEED Humanism		
Chapter 3.	NEED Humanism	39
Chapter 4.	NE: Non-Eurocentric Humanism	57
Chapter 5.	E: Ecological Humanism	87
Chapter 6.	D: Durable Humanism	102
Chapter 7.	What Next? Good Guys and Villains?	122
Chapter 8.	Reset or the Extinction of *Homo Sapiens*?	127
Conclusion. The End, for Now		163
Afterword *Tim Ingold*		165
References		169
Index of Subjects		175
Index of Names		177

FOREWORD

Laura Nader

It is not by recent intellectual detour that Rik Pinxten brings an anthropological perspective to humanism. From the early days of his initial training in philosophy, ethics and religion he has been thinking about what has seemed to him to be missing from the narrow Eurocentric, Enlightenment origins of humanism. Due to its exclusive boundaries, humanism is not a concept commonly used in anthropology today. Rather than reinforcing separate tracks, however, Pinxten proposes that Western contemporary humanists embrace and incorporate elements from the full range of human culture.

Humanism dates from five centuries ago and can be considered deeply rooted if we measure the sheer volume of written thought that underpins it, but, Pinxten argues, it lacks much in terms of a contemporary understanding of humanity itself. Early humanists, such as Montaigne, were more open-minded about what non-European peoples might have brought to a European rethinking of how people might improve their societies. But it was not to be as the narrow canon solidified. In this volume, Pinxten makes the ambitious case that this narrowness can yet be expanded.

The discussion on humanism in Flanders, Belgium illustrates the general development worldwide, focusing on economic and political cosmopolitanism in which Pinxten rethinks concepts like freedom and democracy to challenge right-wing non-humanistic movements that seem blind to a world that is increasingly interdependent. Much of the enduring damage caused by Western expansion over the past five centuries – contemporary global problems such as anthropogenic climate change and ecological destruction – is clearly geographically non-Eurocentric and thus requires a multitude of cultural, economic and political perspectives to begin to restore humanist ideals.

Sustainability is the measure on the agenda today, everywhere. How to define it, measure it and work towards it are central issues to all of humanity – especially in the West given what the grim determination of the Eurocentric worldview has wrought. Humanists confronting these nearly existential challenges would do well to heed the new avenues and frames of thought that Pinxten entices them with in this important book.

Laura Nader is Professor of Anthropology at the University of California, Berkeley. She initiated legal anthropology and has published extensively on law and conflict regulation in other cultures (e.g. Mexico). She is active in speaking out about public issues around law, politics and discrimination.

Preface

When Thomas Mann was writing his famous *The Magic Mountain* about a century ago, he was a visionary. In my reading he tried to picture how the tired and worn-out continental European worldview was giving way to the so-called liberal Anglo-Saxon concept of humanity and community, where individualized short-term relations would transplant the older, slower and so-called less free continental ways.

In the book (see Mann 1996), the Jesuit Naphta lost the final fight against the humanist-freemason Settembrini, but this was not a victory for humanism in Mann's novel, since both the old Church control of the Jesuit and the humanism of his opponent appear as outdated worldviews, deciding on a fight which did not matter anymore in the light of the historical moment (the outbreak of the First World War). In a modest way, the present book refocuses on humanism in a similar perspective: the 'salesman' Anglo-Saxon worldview, I believe, has indeed won the game after the First World War, but its offspring is now threatening to abolish any kind of humanism-cum-democracy. One alternative might be (together with scientific steps forward, of course) to deeply reset the old humanism. In a sense, Settembrini's second and never seriously developed project could be phrased as follows: how can humans redefine their place in a world which they largely co-determine?

When engaging in the translation of an earlier Dutch version of the present book I came across recent works of two of my favourite novelists. To my immense joy they both address similar issues: Richard Powers (2018) calls for horizontal connectedness and Dave Eggers (2021) deals with the question of freedom under surveillance capitalism. Both themes are central in my critical appraisal of humanistic-democratic thinking in and for the present world. This gives me hope: art, science and philosophy are reaching out to one another again.

This book is about values and norms, one could say. However, it deals with such issues in the context of an overzealous and inconsiderate capitalism that steers and uses science and technology more and more for its own profits in the short term. It also threatens to push humanity to a point of no return, maybe even to annihilation of *sapiens*. In discussing these issues the book also engages in a deep analysis of the western appreciation of other cultural traditions. I consider myself to be a humanist and heir of the Enlightenment. In that sense I hold a lot of the social, epistemic and cultural products of what is commonly called 'western civilization' in high esteem. At the same time, I am convinced that this broad humanistic tradition stopped being self-critical decades ago and shimmered away the more a raw and gradually globalizing economic ideology of capitalism took the lead in social and political matters. With so-called neoliberalism this ideology is becoming a real threat to humanity and other species. Moreover, it is stunning how, during the COVID-19 pandemic and with the alarming climate crisis, humanists of various stock can still not find a decent, humane and sustainable answer, thus sending the message to the next generation that 'après nous le déluge' would become the shared rule of conduct. Transhumanism seems to voice that message, but the exclusive rationalism of devoted humanists today does that too.

Cultural Relativism

I am not a cultural relativist: many cultures exist but, as far as we anthropologists know, none of them is deeply or structurally isolated. Anthropologists have shown clearly that intercultural contacts – and, with them, travelling genetic and cultural material – have been with us at least from the time of the Neanderthals. The intrinsic superiority of one particular ethical, religious or political construct over that of neighbours or 'strangers' is, then, impossible to uphold, just as the insular and exclusive superior value of any of the traditions we come to know of appears to be a rather romantic chimera (a sort of Crusoe view, one might call it). The view that 'other cultures' would be better or even essentially different from ours – in the sense that essences or substances can differ from each other in chemistry – has been the basis of cultural relativism in the past centuries. In a broad sense it can be said to be mistaken, just as much as the supremacy view is today.

What we can say is that we, the West (or first of all the European countries, followed for about a century by a set of other states), have, for approximately five centuries now, built and continuously upgraded a forceful instrument to conquer the world. This instrument, I think, consists of a tight complex of knowledge (science and technology), economic and financial structures and attitudes, and military power. Presently, the deployment of this instrument, used generally to extract profits from all over the planet, is arguably steering the whole of humanity on a road to self-destruction. 'They', the other cultural traditions, did not do that; 'we' did. Or at the very least, they did it to a far lesser extent. But maybe they will occupy the leader's position soon (as some would claim with the rise to power of China, lately). But whatever the future will bring, we do not have any seriously substantiated, 'proven' or well-argued reason to claim that any one cultural tradition in the world has a markedly superior ethical, religious or political standard to any other, in a generalized way. If we grant this point, this implies that arguments to defend supremacy cannot be used by us, westerners, to continue on our track and refrain from self-critically looking at our values and norms. Having said that much, I will claim that we should decolonize our views on other traditions and on ourselves (as the superior culture), and critically assess and reformulate even deeply rooted concepts and principles. In that vein, the old concepts of humanism, worked out further throughout modernity and Enlightenment, should be looked at carefully.

On top of that, when nothing short of a 'coup d'état' almost took place on 6 January 2021 in Washington, DC, this might be better seen as a wake-up call to humanists/democrats and invite all of us to scrutinize the principles we took for granted in a world that might have changed profoundly over the past few generations. Indeed, Washington was no unique accident, one might argue: the success of extreme-right parties in 'established' democracies such as Italy, France, the Netherlands or Belgium in recent years points to the same doubts in large parts of the population about the sense of the humanism-Enlightenment project. One question could be: why and where did the so-called mob in Washington, DC stop endorsing this humanistic tradition? And secondly, and most importantly, in what sense could and did humanism speak of and for humanity in general, when even groups within our midst openly rejected this programme in an armed attack on one of the world's main instantiations of it, or during recent elections in several European countries?

I mention all this to make clear that such self-critical analysis has nothing to do with cultural relativism. As an anthropologist I hope to be able to offer some clarifications here: what is culture all about, and do westerners differ from the next 'culture' or is humanity diverse in another way? And what impact does all this have on an open, self-critical humanism? After a systematic analysis I will sum up these questions again (chapter 7).

Preface • xiii

Common Interest and Private Interest

In my view we are witnessing in this age yet one more interesting switch: what used to be understood as private interest (*'it is my right'*, *'it is my private property'*, etc.) starts to collide in crucial areas with common or global interests. For one thing, the present COVID pandemic showed the way to all who are willing to see: due to the global mobility of people and goods 'what happens here happens everywhere'. A health problem of this nature spreads all over the earth in no time, just as economic problems (like the shortage of microchips after the pandemic, or the crisis in energy) are no longer local, but ring throughout the globalized world. Moreover, it becomes clear that the more we stick to private or local interests the more we risk unsettling processes and structures on a global scale, causing conflicts, destabilization ... and streams of refugees. Those looking at climate change for the past few decades already knew about this, of course, but the general public probably only became aware of it with the pandemic. Finally, the intense interrelationships between climate change, destruction of biodiversity and recent structural inequality in the world (to name just three processes) were known by broadly interested scientists and intellectuals, but broader consciousness of these issues is likely to grow now – also occasioned by the COVID pandemic. Speaking quite generally, I can say that the high value of independence and exclusive sovereignty, which was held to be almost sacred in the past two centuries of nation-state building (also sharpened through the wars between states), is rapidly becoming questioned: for the very survival of any and all in an economically and ecologically globalized world this political frame may be largely dropped out of necessity, at least for some domains, to be replaced by the value of *interdependence*. For the survival of *sapiens* interdependence is a rapidly growing reality, a factual given. However, our value systems and ideologies are still largely stuck in the outdated view of independence, as exemplified in the state institutions. My intuition is that humanity will probably not survive (or will at least experience tremendous losses) unless we all start reasoning in togetherness, recognizing that the fate of anyone is tied to that of all of us on a global scale. That is the case for climate change, but also for inequality or poverty and for a healthy survival of humanity in general. Hence, the isolationism and cultural-identity ideologies that we see emerging here and there (and of which the Washington coup may have been an example) are either outdated already or calling for extremely violent conflicts between the haves and the have-nots as a kind of outdated attempts to withdraw to one's isolated community, much like an island of survivors. However, when I start analysing in a critical way the tenets of humanism, as the foundation of many (western) democratic institutions, I find very little concrete plans that refer to earth-wide interdependence. For instance, while factual interdependence is becoming obvious, a redefinition of local and so relatively mutually autonomous 'circular economies' may become more important for our survival. There is no contradiction here, but a deeper understanding of the whole will be needed – which implies that old humanist tenets need reconsidering.

Many will most likely retort: but most western citizens have a good life, don't they? And many politicians and philosophers will explain that our wealth is caused by our superior values and our tradition of working hard. So where is the problem? To be sure, during the so-called 'heureuse trentième' ('the happy three decades' of 1950–80: Piketty 2014) the politics of redistribution in many western countries made life in these countries better for many: life expectancy rose for most, medical care ameliorated sickness, children got more opportunities through a better educational system, and social security was generalized in several countries for the first time in history. Inequality diminished during this brief period after the war (Ibid.). Of course, this was made possible here to a large extent by an enhanced and even more systematic extraction of wealth in other parts of the world: resources from poor countries were systematically taken from there at low cost, cheap labour was found there and waste was massively exported to these regions (or dumped in oceans).

The well-meaning westerner will point out to me that the rights of women and even of children were enhanced in the west (but not so elsewhere), amongst other positive developments. The heavy debates on such issues (see #MeToo) show that the battle is not won yet most of the time, and that progress tends to be local and sometimes at the expense of people in other parts of the world. I suggest looking at such arguments in some detail, but also within the general frame that is ours today: I claim that we have to rephrase the question deeply so that the interests of me or my small part of humanity will be seen from now on in the light of the interests of all. Our fates are not just 'somewhat' linked, but are intrinsically tied together. That is the focus in this book. When and where does humanism grant this point – and if not, how can we go about the reset necessary for it to do so?

Global Interdependence

As hinted at in the former paragraph the reckless kind of capitalism that has been spreading over the world, especially since the twentieth century, is threatening the survival of humanity. The control mechanisms and institutions, especially the United Nations (UN) after the Second World War, are relatively powerless in these times. Of course, the Intergovernmental Panel on Climate Change (IPCC) emanated from this international organization, but the final decisions are still made by leaders of states who gather regularly in so-called climate forums with government representatives: so far, decisions there have been insufficient, aside from the fact that their implementation through national policies falls terribly short of the promises made at the international meetings. Unfortunately, COP26 (the 26th UN Climate Change Conference of the Parties in 2021) did not show any different, although every participant was aware that time is extremely pressing now. Even in the case of a directly recognizable and in itself manageable crisis like the COVID pandemic, honesty forces any observer to recognize that the rich countries invested massively in vaccines but the poor countries were mostly left to solve the problem on their own or to be happy with occasional leftovers from the tables of the rich – which yielded bleak results because the local and private patent law of Big Pharma in the west would largely bar access to the remedy for the poor. Again, the sense of interdependence is lacking although its factual presence is now obvious to all. In the case of vaccines, the international agreements on patents and private property hinder efficient panhuman thinking and acting. What is humanistic about these practices, knowing that the bulk of scientists and industrialists involved were raised in that supposedly superior humanistic tradition of the west?

Still, it all began in a rather promising way. After many centuries of domination of European regions by the 'religions of the book' (roughly from the sixth to the fifteenth century) – with their political, educational and cultural uniformizing of the minds of the population of the continent and the islands – a critical reaction began to grow in the minds of what we would now call the intellectuals of the fourteenth–sixteenth centuries. Erasmus, Giordano Bruno, Giambattista Vico, Thomas More and Michel de Montaigne attacked the heteronomy of the religious view with a proposal of greater human autonomy. Heteronomy held that ethical, political and even cognitive solutions and choices were thought through by the Creator, who then revealed them in his Will, laid down in the words of the Bible (with the Scriptures and the Qur'an in the two later religions of the book): the authority of these texts should and could not be doubted or denied by any of the creatures, i.e. the human beings. The answers to all questions in a deep sense came from outside (that is, hetero-) the humans themselves, namely from God the Creator. The early humanists started to doubt the power and the authority of these texts. They developed the view that human beings themselves could make things out and decide about moral and political issues on their own, in the interest of humanity. Humans became the measuring rod of things; autonomy would henceforth become highly valued in several domains, such as knowledge. Free thinking (to some extent at least) proved more sensible than ancient texts, which did not easily relate to the changing perspectives on reality (early science, but also the beginning exploration of the world outside of Europe). With the adoption of scientific methods later on (i.e. with the Enlightenment and the foundation of scientific academies) this trend towards self-determination of humans in cognitive, ethical and political questions only grew more powerful, thus making this humanistic turnover in Europe (and some colonies like North America or Australia) the dominant perspective in several countries. In actual fact, differences within the West remained: secularization is the baseline in France but not nearly so in the USA, for example.

Of course, humanists were people of their time and cultural background: most of them would keep reasoning within a racist frame of mind (as we would call it today): whatever was found in the world that differed sharply from the European perspective would be seen as primitive, or even part of 'nature' and hence not of humankind. From that point on it was easy to disqualify it as inferior rather than merely different. Exceptions can be mentioned (notably Montaigne, who showed interest in and respect for members of other cultures), but most humanists and later Enlightenment thinkers in Europe and North America stuck to the deeper Christian division of the saved and therefore superior humans (i.e. the followers of the religions of the book in general or Christianity in particular) and the heathens. This explains why so-called enlightened thinkers like Voltaire, for example, could have a comfortable life on what they earned from the slave trade of their time. There must have been no serious contradiction in their minds between humanism and Enlightenment on the one hand and the qualitative distinction between peoples or 'races'.

Humanistic–Democratic or Not

The humanistic project, which started about five centuries ago, had a glorious short history. After the first negative reactions by established churches (especially the Roman Catholic Church, which burned people like Giordano Bruno at the stake and tried many others), European education and politics to a large extent took this view on autonomy of human ethical judgment seriously and started organizing research (with the scientific method), and to some extent the political and social structuring of communities, along more humanistic lines. However, in the present era I see trends in the west which claim that humanism is outdated, or – in terms of power relations – 'unrealistic', and hence should be left behind. Conspiracy theories, the strengthening of heteronomous ethics (through more fundamentalist interpretations of religious caveats in all of the Mediterranean religions) and the rejection of scientific authority are growing. For many 'old' humanists, these processes appear odd and ill-conceived. For me, such an attitude of denial fails to recognize serious problems stemming from the context wherein humanism developed. And, I claim, humanists have done too little to think through the contextual restrictions and demands that we are all subjected to. I list a few points here.

Scientific knowledge has been explicitly questioned in some powerful western countries, heirs of the humanistic tradition: the recent claim of 'alternative facts' and the explicit doctrine of 'post-truth' in official political strategies of the USA, Brazil, India, Poland and Hungary (to mention only a few) make clear that the old humanistic stands are becoming less popular. Of course, intellectual discussions of whether or not, or to what extent, fascism and certainly Stalinism and other branches of communism must be considered heirs of humanism (and of the Enlightenment in its slipstream) yield odd but certainly not simple questions as to whether concentration camps can be conceived of without the outgrowth of self-determination, planning and scientific rigour that sprouted with humanism (e.g. in the critiques of some postmodernist philosophers).

Finally, the coupling of the organization of scientific and technological research on the one hand with the development to capitalist economic policies on the other gradually broke some of the humanistic claims down to manageable and marketable formats. For example, since the Fall of the Berlin Wall (1989) researchers throughout Europe and the USA have had to 'generate output', raise research money and show what patents their research will produce. At the same time, university staff are engaged in extreme competition in the short term with each other, while managers are organizing a very systematic market competition amongst researchers.

Lately, more and more universities and research institutes are hiring CEOs from corporations to 'run their business', which has led a major anthropologist and critic of the disappearance of free intellectual thinking in the west, Marshall Sahlins (2008), to speak about the 'corporatization of academia'. Instead of God's law the so-called laws of the market will dictate what questions to ask and what type of answers to take seriously. This impact of economic values and interests over free thinking and self-determination could not be recognized by the mainstream humanists in the past, but neither is it their principal focus in the present. In past decades, for example, the more conservative humanists even started separate institutions which today deny that economic incentives or even power interventions are genuinely relevant to the humanist. Their focus is exclusively on thinking 'free from religious authority'. For instance, in the Netherlands, representatives of this line of thought distanced themselves from the mother institution; not surprisingly, they can be found today in rightist political parties (such as the Forum for Democracy) which tend to be anti-Islam and anti-socialist. In Belgium such a tendency can be found in Flanders (without institutional splits so far), but much less in the French-speaking parts of the country. In France, the extreme-right party of the Le Pen family (the old Front National) claimed at a certain time to represent the real French secular tradition, which yields anti-Islamic and anti-socialist stances. Thus, the impact of the political–economic context in which humanism did and does develop is at the very least insufficiently accounted for.

What about the Anthropological Turn? And the Ontological Turn?

For a few decades now, anthropology has been changing its perspectives: after empiricism and its uneasy lack of critique of the colonial attitude of the times, the Durkheim–Mauss–Lévi-Strauss perspective made clear that researchers are not blank sheets of paper on which data fall and which then produce imprints of an otherwise virginal external reality. The discussion between the empiricists (led by Alfred Radcliffe-Brown in the UK) and the 'deep' structuralists (led by Claude Lévi-Strauss) has in a sense never ended. Today we find intense and in fact deepening debate on the nature and impact of anthropological knowledge, I think, linked to what is often called the 'ontological turn'. And this lands us right in the middle of the question of whether anthropology may be changing internally to such an extent that it often takes the place that was occupied by philosophy in the past centuries. Put differently, the shift can be characterized as a possible anthro-pologizing of domains that were for a long time the sole privilege of philoso-phers. To name just a few scholars in this evolution: Johannes Fabian (1979) already makes this point, after the older analyses by Claude Lévi-Strauss (starting in the first volume of *Anthropologie structurale*, 1958). In the past few decades we can point to the anthropologically inspired philosophy of Jacques Derrida, but also the influential work of anthropologist–sociologist Pierre Bourdieu, the impact in many disciplines of work by Clifford Geertz Marshall Sahlins or, lately, economic anthropologists such as 'anarchist' David Graeber or 'degrowth' thinker Jason Hickel. Over the same half century or so colonies have ceased to exist, but a neo-colonial mentality became more visible, and the reaction against it is reshaping public and academic debates to such an extent that race, gender and cultural identity sometimes occupy public space almost completely at the expense of class struggle, poverty and indeed interdependence (as in the climate crisis, the crisis in biodiversity or the new inequalities in the world of this century). Obviously, topics like race, gender and suchlike are treated importantly by social scientists, and among them prominently by anthropologists.

The net result is that a deep discussion on diverse 'ontologies' (emanating from the Indigenous American scene, primarily) instead of THE ontology, developed over the ages by armchair thinkers–philosophers, is often fed by anthropology. In a recent discussion on such issues by two leading contemporary anthropologists of the west (Tim Ingold from the UK and Philippe Descola from France), the continuation and critical assessment of a possible turn (ontological and/or anthropological) was central (Ingold 2016; Descola 2016). All this will be dealt with in more detail in several chapters of this book. It is mentioned here since, in my view, humanism needs to be detached from its exclusively western philosophical background and fertilized by contemporary anthropological ideas.

I am grateful to colleagues and friends who read and commented on former versions of this text, leading up to a book in Dutch (Pinxten 2021), which I now consider to be the forerunner of the present volume. I thank Gily Coene, Johann Leuckx, Marianne Marchand, Willem Meijer and Erika Vercammen. They reacted when I was writing in isolation during the COVID pandemic. Of course, my wife, Ellen, and the children gave sense to this one specimen of *homo sapiens* in his endless attempts to try and understand what the meaning of life could be.

NOTE ON TEXT
Warning

Having reached in all probability the last quarter of my life, I think I can say that I am a humanist of sorts. Of course, the term covers everything and nothing. In my case, and probably somewhat differently from fellow humanists, I firmly hold that humans are not the only species with communication (language of some sort) and a social life. Trees are joining this club, present-day botanists tell me. Neither are we the only species that thinks, feels or has something called emotions. It might even turn out that we are not unique in having beliefs, habits and a will. That our will can be free, as we started to believe a few generations ago, is debatable to a considerable degree and depends heavily on what this notion of freedom can come down to anyhow. Whether we are capable of taking up any of these notions of freedom is, certainly in an era of 'surveillance capitalism' and Chinese obedience society, at the very least a question, not a safe answer.

After a life of studying cultures and religions on this earth, I am convinced that humans, as a species, have this one unique characteristic (or quality, if you like) – that they are able to lie or invent phantasies and then adapt their beliefs and behaviour accordingly. Put differently, we humans can formulate criticism on what appears as normal (the norm) and look for something else that does not yet exist. We are not the only ones who lie/make up things: animals play and hence 'redefine' elements in their environment and even on their bodies. But what humans seem to be able to do is make up parts of their world by adding impossible things (levitating elephants as in Eric Peters' paintings, beauty, etc.) out of nothing, or referring to no known or experienced reality, and then agree that their lives should now be reconceptualized and ethically and socially reorganized according to rules, sayings or the 'will' of these phantasy creatures. That is shown in all forms of creativity that this species is capable of: in religion/ life stance or worldview, in art, in knowledge, and in societal or political models. We are probably the only species that does so. And that double-sided capacity is the basis for my view on humanism, in a nutshell.

In line with that, I am inviting the reader of this book to engage in a deep reset of the known or 'established' humanism-cum-Enlightenment programme that was so typical and successful over the past five centuries, in the conquest of the world by the West.

I speak in this book about humanism-cum-Enlightenment in one complex conceptual cluster. Of course, there are many differentiations, but it is not my aim to have a deep understanding of the relationships between humanism and Enlightenment as historical phenomena. Other studies and even research centres do that (e.g. the Centre for the Study of the Enlightenment at the VUB University in Brussels, Belgium and the University of Humanistics in Utrecht, the Netherlands). For the purpose of this book and in the comparative view of the anthropologist on cultural differences and life-stance diversity throughout the world (see Philippe Descola's insistence on these points: e.g. Descola 2005), I see one basically shared set of principles that runs from humanism in the Renaissance till today. In that sense, I can easily agree with Claude Lévi-Strauss's saying that Michel de Montaigne is one of the main philosophers in Western philosophy, although the latter did not develop a 'system' like Enlightenment philosophers such as Immanuel Kant (Lévi-Strauss 2016). Montaigne started a programme of questioning received views in a way that can still be considered to be revolutionary today, through modernity and even our present era: his contemporaries (like Erasmus, Giambattista Vico, etc.) were less 'radical' and stayed more loyal to many intuitions and principles than he did. For example, they did not question as deeply the Eurocentrism, nor the identity-cluster of religion as a set of beliefs, as Montaigne did. His successors shied away as well when they translated a lot of the humanists' critiques into political and economic frameworks and models throughout what is called the Enlightenment. Montaigne seems to have been able to focus on questions and questions only, and that may be precisely what we need in the present era. Again, I am conscious that I generalize. I should make an exception for some thinkers (notably Diderot), but looking back from the present state of the world my point remains valid, I claim.

I propose to look at the intellectual complex which was developed in Europe throughout the past five centuries as one stream, disregarding the many internal differentiations that can be spotted. I claim it is relevant and important to see the relative unity here, when looked at from the point of view of a comparative anthropologist. With the deeply disconcerting self-critical re-evaluations that are offered today in the works of, for example,

Philippe Descola and David Graeber, I claim that it will prove healthy to look at our tradition in this way. It will prove healthy, notably, when we start the reset of our way of dealing with fellow humans from so-called 'other cultures' and with nature. That is the invitation I send out with this book.

Introduction
The Zebra and the Dolphin in Us

What Went Before

The Second World War seems to have taken most older people whom I came to know in my life as if by surprise. Or, better, they spoke about it as a terrible conflict of and by other people, over their heads, probably about money and power, but not really about themselves. I imagine that those who joined the fascists and those who fought them in the resistance were engaged in another way, but the so-called mainstream people I knew did not position themselves politically with either side. My father was born during the First World War, in Antwerp, Belgium, in a low class in a so-called popular neighbourhood. He was the late Benjamin of a small bakery, and so he was delivering bread in part of the city early on, from the age of seven, driving a dogcart before and after daily school hours. By the time he was starting primary school his mother had died of pneumonia, the raging lung disease that also killed his elder sister only a decade later. My paternal grandfather, the baker, turned alcoholic after the death of his wife and his only daughter, so my father told me many times about his adventures as a child: he had to go and search for his sad father in one of the many cafés where he landed after his round of delivering bread. At the end of each day, my father was then to feed the dogs of the carts, an activity he told us about until his old age. Although he was a clever pupil, his school results were meagre. During puberty he flunked at school and went to help out as an apprentice of sorts in the plumber's shop of a distant relative. After a couple of years he decided to have a go at it and start out as a plumber on his own. But by then the Second World War had started; he was drafted almost on the eve of his marriage, only to come back home a year later. For almost

2 • Humanism Revisited

five years life froze for the young couple in the city of Antwerp, where occupation and heavy bombing were the rule. After the war Flanders in general, and Antwerp in particular, was reconstructed and basically industrialized for the second time in its history (after the glorious sixteenth century and the brutal destruction by the fundamentalist Christian king Philip II in 1565), in a rapid way. During that period of reconstruction under the Marshall Plan, we, the three sons, were born.

My mother was a clever woman, coming from similar low-class background. She had lost her mother a few days after birth. Her father had promptly remarried: he took a sister of the deceased mother as his second wife, and my mother (rightfully, I think) sometimes had me understand that her stepmother had never been fully able to love her. The stepmother preferred her own two sons. At the age of twelve, my mother was sent to work as a tailor-help in a large shop, where she sat on top of a long table with a dozen other young women for ten hours a day, sewing costumes for well-to-do customers. Her intellectual needs were only very gradually fed, when she reached older age and started reading literature and philosophy all by herself. I learned from her that, notwithstanding the need and the tremendous effort one puts into such an engagement, a lack of guidance or education is extremely difficult to overcome on your own.

When I did research on the children (in their school context) of the same neighbourhood more than a generation later, as the head of a government project in Flanders dealing with intercultural education and serving many primary schools throughout Flanders for about three decades, I often had moments of what looked like a flashback. Yes, times had changed, but the so-called 'newcomers' in the old neighbourhoods – mostly immigrants and their children – lived very much in the way I had been doing in my early years. Where my father by chance survived certain predicaments in his youth, regardless of a lack of interest in and facilities for difficult or disadvantaged children in his day, I witnessed during my youth in the same neighbourhood that out my class of twenty-five pupils in elementary school no less than five ended badly. One ran his 'upscaled' first motorcycle into a lorry and died in the accident at the age of fourteen. Two others were caught on the verge of raping a girl they had tied down, in a little marketplace nearby. One boy ended up living on the street, and one – the only bourgeois boy in the group – went from depression to depression throughout his life, utterly incapable of meeting the standards of his family. When, as a researcher, I looked at the profiles and school careers of the children who lived in the same neighbourhood today, the same percentage of dropout can be found. The only difference might be that today's children of the new residents are overwhelmingly those with a migrant background. Indeed, the neighbourhood of my childhood changed little, except that the population was getting culturally (or ethnically) mixed. But poverty reigned just as well, while the so-called socio-political problem was still identified as a problem in the children and their parents, rather than in the segregating policies they had to live under. The obvious conclusion I drew, and still draw, is that the policies that have been made up and applied in schooling and neighbourhood management are inadequate. This is, in other words, not a problem that can be identified with, let alone blamed on, the poor people living in that part of town, but clearly a political problem. Put differently, it is not the people's private opinions and values *as individuals* that will adequately work to change the perspective on life, like the old humanists pretended. Adequate and hence humanly correct and fair policies will have to deal in a courageous way with the people in their context(s): their economic, cultural, social and ethnic contexts. In order to be humanly correct and honest, I think, one has to take humans-in-context into account, and that cannot be done in an a-political way.

Having said that much so far, I can refer to my first book on the issue of humanism. About fifteen years ago I was asked to occupy the so-called honorary chair Willy Callewaert at the Free University of Brussels. This chair aims to promote fresh thinking on humanism in this part of the world. In my lectures I emphasized some of the issues I have related in the former paragraph, referring to them as the 'stripes of the zebra' for humanism (Pinxten 2007). This metaphor needs some explication. Anybody recognizes a zebra immediately, distinguishing it without fault from other species of the horse family. Any zebra is a zebra, because it has stripes all over its body. On the other hand, every individual zebra has a particular set of stripes, distinguishing that animal from all others. They have stripes as a species-specific feature, but the patterns of stripes are typical for one particular animal. Working with the metaphor, my plea in the lectures and the book was that humanists should be able to see and advocate that something similar obtains for humans: the species is human, but within the species a wide diversity of features obtains. We have a definite degree of zebra-hood, and the old humanism did not recognize this. This was mainly because it developed concepts and models within one part of the species (i.e. the European, mainly and deeply exclusive Christian context) and 'universalized' the features of this subgroup to the rest of humanity without sufficient awareness of the diversity.

Excursion: Difference and Diversity

The Oxford Dictionary teaches us that difference is the recognition of 'unlikeness' of one entity, object, process, etc. vis-à-vis another one.

Diversity, on the other hand, focuses on multitude or being many (aspects, facets) within one and the same phenomenon. Obviously, the colours white and blue are different from each other. However, they are values of diversity of the one domain/phenomenon that we call colour. When speaking about human beings the present-day fashion of difference-thinking, aka identity thinking, claims that human communities (or eventually even individual persons) can be understood as 'us' and as essentially other than 'them'. Hence, we can justify thinking of 'us' against 'them' even when both groups are part of humanity. Particularly since the 1990s (with Samuel Huntington and others) the us–them difference has been promoted continuously, adding that this will inexorably produce 'clashes' (as the title of Huntington's 1996 book suggested). Cultures (understood by these scholars as groups or another social set of persons with particular types of relations between them), races, religious communities and suchlike will then be considered distinct 'entities' which together will somehow form humanity or the human species – hence speaking about each entity in the whole as 'different' from the next one. The alternative view is to see the whole, humanity, in terms of diversity: diversity is an intrinsic characteristic of one 'entity' or biological, demographic set, within which nuances, shades, more particular features can be discerned at a more superficial level.

With a rather risky metaphor one could think of mayonnaise: essentialists who side with the difference view hold that there are basically two essences which are extremely hard to mix in order to become the emulsion wanted: there is oil and there is vinegar. The mix happens only under very strict rules, neutralizing the difference to a sufficient degree. The diversity perspective holds that the new substance of 'mayonnaise' can be discerned in a variety of forms and shades, all located in a broad range of the one substance: rather more or less sour, lighter or darker in colour, and so on. But all of it is mayonnaise consisting of the same basic ingredients in a continuum of values. Dividing this whole as if different 'essences' (in a chemical sense) could be identified as inadequate, since it demotivates the search for the delicious new product.

The emphasis on the terminology is not trivial, though. A quick look at the history of chemistry should make us aware of the influence of cultural and political biases: a colleague in the history of science told me the story of the beginnings of chemistry when the so-called phlogiston controversy was raging in France and England (at the end of the eighteenth century, with phlogiston as a forerunner of the oxygen element). Antoine Lavoisier and Joseph Priestley were heavily debating the sense or nonsense of the distinction of two types of phlogiston, namely a 'male' and a 'female' entity to their minds. Although at some point everybody agreed that this difference was not only right but also justified (since the male one was clearly bigger and weightier, and the female one less imposing, thus being in line with Christian views on gender difference), the discussion was resolved once both camps learned to drop their ideologically based way of looking and recognize that oxygen behaved in diverse ways when linked to other elements. Now we all agree that the science of chemistry was really only launched once this ideological type of essentialist thinking was dropped.

The other way of looking at humanity is that of the natural scientist: there is one species, *homo sapiens*, living on the earth for the past two to three thousand centuries. We came to understand recently that this species carries genetic material from Neanderthal predecessors, for instance (Condoni and Savatier 2019; and Nobel laureate Pääbo 2014). This indicates that biologically diversity is an undeniable fact for the present species of humans and 'difference' can, if at all, be only shallow.

One more recent argument should be mentioned as well. Contemporary genetic research also showed that human groups were never isolated in a deep or intrinsic way: groups traded with each other throughout the world, they migrated, they waged war on each other and they had offspring with 'neighbours' from time immemorial (see again the genetic studies of Pääbo 2014). Given the year-round fertility of women during several decades of their life, the way in which genetic material (like cultural forms and objects) is involved in continuous streams or travel routes over the globe can be measured. Thus, John Relethford (2006) calculates that any genetic mutation occurring, say, in the most southern part of South America will be found within a period of minimum three hundred and maximum three thousand generations in the north of Norway or in Siberia. Genuinely isolated 'cultures' are therefore a fiction. Communities could live in an isolated way for generations, but not in the sense that they would never belong to the one interconnected species of *sapiens*. In that sense diversity is a feature of humanity, and difference is most likely a temporary cultural interpretation of certain parts of humanity.

This excursion wants to highlight the relative relevance of both concepts. At the species level only diversity can be recognized, with a bit more or a bit less Neanderthal in one and the same species, for example. When we emphasize difference – as in identity politics, 'clash of civilization' thinking or the religious exclusion of variants of the presumed, unique, true version – we lack any deep foundation on differences, like the genetic base we have for biological knowledge. How deep or even genuine are historical differences, when we know that just after the start of the *sapiens* species mixture with the Neanderthal species (and not extinction or absolute exclusion) occurred? And what about cultural differences, when we know that genuinely isolated cultural communities are a fiction? Rather, we should start thinking and speaking in a responsible way about degrees of diversity within the one species of *sapiens* and thus start negotiating ways of communicating and interacting between diverse cultural survival forms, now locked together more than ever, in an englobing relationship of interdependence. The question then becomes: what kind of humanism can figure in that sort of world – and how?

The Dolphin and the Zebra

Gradually, I came to feel uncomfortable about my one-sided emphasis on the zebra-hood of humanity. Not that it was wrong, but rather that it was incomplete: it lacked an important dimension, and hence was understood too easily in just one sense. The present pandemic, again, shed some light on the question. Indeed, as a culmination of processes that started earlier, we drifted into the world of crisis that we lived for at least two years from 2019 on a global scale. Parallel to this modern version of the Plague, it became extremely clear over the past two or three years that poverty and general inequality had been growing ever faster since the neoliberal free-market ideology came to power in most industrialized countries in the world (normally dated from the 1980s, the coming to power of Margaret Thatcher, Ronald Reagan, etc.). The following, not limitative, series of events struck me:

- The West was very successful in identifying the virus and developing, at unforeseen speed, effective vaccines. Together with the rather well-organized healthcare system in most of the wealthy countries the population was protected against an extremely devastating spread of the virus.
- On the downside, I saw a rush for the vaccines which was won by the rich countries, at the expense of vast parts of humanity. The will to recognize interdependence was still weak, leading to the awkward and despicable situation that private ownership (of patents) would win out over solidarity, including that one would be saved from a horrible death in rich countries but not so in poor countries.
- On the positive side, I witnessed a remarkable willingness in stacks of common citizens to help each other: solidarity was not dead, as the prophets of neoliberalism would have preferred, and also ordinary people largely manifested their will to respect rules in order to have the community survive.
- On the negative side again, we saw conspiracy thinking and 'alternative fact messages' spreading in unforeseen ways through the new means of communication, the offspring of internet technology. Extreme-right movements and religious-sectarian denominations found each other in this time of fear (Höhne and Meireis 2020).

- On the positive side though, governments in democratic countries and elsewhere were given the authority to govern actively and firmly in order to beat the pandemic: notwithstanding years of neoliberal criticism of too much government or state, overnight so-called 'essential' sectors and activities were distinguished from superfluous ones. Healthcare, education, law and order, food and public transportation were recognized as essential for the survival of a community and firmly steered by governments, whereas stock-market activities, corporate meetings, sports or luxury shopping were largely shut down by the same in an attempt to control the spread of the pandemic. Cultural events were, amongst some others, undecided. People quite generally agreed on this ruling and obeyed the rules from their government (at least in the first year or so).

When I look at this balance I conclude that the dolphin-hood of humanity may have won, at least 'on points' – and hence I want to explain this feature a little more: empathy was shown to really count, I suggest.

Dolphins are a peculiar species. They have developed a rather elaborate language (of some twenty-one distinct sounds), raise their young in social units and show the unique quality (in the world of more sophisticated mammals, that is) of helping the members of their species and even those of other species when they get in trouble. This unique quality was remarked on by early western seafarers who tried to get through the then-uncharted Strait of Magellan in the seventeenth century. They were apparently helped to get through the dangerously rocky narrow strait by these animals, living in these waters. In a nutshell: dolphins seemed to show empathy.

My contention now is that human beings have this quality also, possibly even in contrast to other species such as bonobos. Comparative research on human and animal behaviour at the Max Planck Institute, led by Michael Tomasello (2009), shows clearly that infants develop this capacity from the age of eighteen months on. It distinguishes humans quite clearly from many other mammals, who either lack that quality or develop it to a lesser extent (like the said bonobos, for instance – see ibid.). However, dolphins show remarkable similarities with humans on this point.

It is then all the more remarkable that late capitalism has been promoting the opposite mentality for the past several decades: the neoliberal ideology of late hails egoism, with the odd claim that 'greed is good'. Also, the presumed mechanism of 'trickle down', neoliberals claim, would compensate for the lack of sharing of wealth with society's poorer groups. However, critical economists have shown time and again that the trickle-down promise is a fiction in the present-day market systems (Stiglitz 2011; Piketty 2019). In other words, it would not replace the empathy which is so typical of solidarity and so foreign to egoistic competition.

In my metaphor: the ideal for neoliberal ideologues would be to strive to curtail or even forbid the dolphin qualities in humans in order to make more room for what used to be called 'the law of the jungle'. Humanism – in as far as it focuses on individual freedom first, even at the expense sharing and solidarity, and thus becomes compatible with neoliberal ideology – does not take this dolphin quality seriously in its view of humanity.

In my critical appraisal of humanism the present development in the 'free west', which yielded a generation of hyper-individualism at the expense of solidarity with other humans and with nature, needs to be critically assessed: I am not convinced that the mere principle of 'man is the measuring rod' or that of individual conscientious decision making has been adequate in preventing this new form of raw capitalism, allowing for indifference towards or even straightforward pillaging of the earth and the impoverishment of most of humanity for the benefit of the few. The rapid development of a small group of billionaires, refusing to share their privileges with the rest of society (and creating tax havens) and promoting their anti-solidarity ideology over the past three to four decades, testifies to the fact that this type of self-determination and hyper-individualism grew in the same cultural bedding and must trigger humanists and later-Enlightenment philosophers to think critically about the tenets of their powerful historical tradition. We are in need of a reset, I claim. In a period when democracy is narrowed down to the defence of the privileges of some, in their capacity of individual deciders, it is important for humanists to look self-critically at the noble philosophy which helped to get rid of heteronomous control only five centuries ago. Otherwise, the non-human overpowering authority of a God will have been replaced by that of a select group of human 'haves', manipulating and even creating markets for their private benefit. Yes, they are humans alright, but the same lack of freedom and the same precarious life is still the rule for the large majority (to say nothing of the threats to other species). If humanism did not foresee and has no decent answer to this, then a reset is called for. It is in that light that, next to the zebra-hood, we have to recognize the sense and scope of the dolphin-hood in humans: living with the positive values of diversity and of empathy should be intrinsic to humanism.

PART I

The State of Things

Chapter 1

HUMANISM TODAY

What Are We Doing, and What Is the Context?

History and Standpoints

Some of my colleagues at Ghent University and at the Free University of Brussels in Belgium have done substantial study work on the historical books and treatises on humanism. I mention Ronald Commers and Leopold Flam as the older generation, and Hans Alma, Marc Vanden Bossche and Koen Raes as younger scholars. Also, after the Second World War, philosophers like Leo Apostel and Jaap Kruithof started building on these historical foundations. Their work laid the foundations of an MA degree in Moral Science at both universities, educating hundreds of teachers for elementary and secondary schools and serving as a host of research at colleges and universities in Flanders, which changed the way a larger population learned to think about the meaning of life in this region in the secular era. The general perspective for this degree is that philosophical analysis has to be systematically enlarged by a multi- and interdisciplinary study of humanity, combining law studies, psychology, biology and economics with anthropology and religious studies in teaching and academic research. In that way, this rather unique project (which carries on today, half a century after its launch) uses humanistic foundations and tries to merge them with a broad scientific frame of reference (Commers 2009, two volumes).

In his visionary overview Kruithof (2001) summarized historic humanism, starting with the late Renaissance in Europe and spanning the Enlightenment and the American and French Revolutions to our own day. I mention a few principles: humans are thought of as autonomous beings,

12 • Humanism Revisited

gifted with reason and by that means able to develop as full persons. In the process, the societal context can be transformed in the sense that human dignity becomes a supreme value and superior goal. Against the background of a religiously determined universe in Christian Europe, with emphasis on humbleness and submission (especially during the dominant reign of Roman Christianity, between the sixth and the sixteenth century AD), humanism was launched as an intellectual movement of liberation.

Central concepts were:

- humanity and society are manmade and not given in any definitive way by a creator of the universe and humankind. This is often captured by the expression: 'human beings are the measure of things';
- recognition of the constructive role of doubt, an attitude that allows us to enlarge both knowledge and the happiness of humanity;
- freedom from relations of power, determined by Church and gentry; and
- tolerance as a high value in humans and in society, linked with a belief in the possible impact of doubt and criticism.

This led to positive developments in Europe, which can be attributed to the humanistic philosophy of 'man' till the present day: the elimination of some forms of elitist politics by nobility and Church as expressed in the generalized right to vote, the fight against irrational beliefs and conceits, and the first attempts to think through forms of socialism/pluralism/modern democracy and human rights.

But at the same time Kruithof mentions some negative by-products of humanism in its historical practice:

- An 'arrogant' humanism led to convictions that human beings can claim supremacy over other species, while staying in line with the dualism of the Old Testament (see below, 'Humanism and Universalism'). An uncritical optimism about human capabilities saw human beings turn into reckless predators, who could lay claim on the surrounding natural environment without any limit. Commers (1982) mentioned that the freedom of thought and choice in humanism brought forth great opportunities, but also occasioned derailments because of a lack of caution and respect vis-à-vis other phenomena on earth.
- The idea of progress, which has spread all over Europe since the sixteenth century, certainly produced wellbeing and the enhancement of living conditions there. At the same time, says Kruithof, this idea of progress was gradually becoming an unlimited and quite coercive ideological frame for action (yielding to neoliberalism in the last quarter of the twentieth century). Criticism of it became rather powerless.

- Finally, in the recent trend of transhumanism a new elitist type of thinking and acting emerges as a power construct by a technological elite, which capitalizes on the progress issue for its exclusive interests, bypassing the focus on humans and on human global wellbeing. With astonishing speed 'surveillance capitalism' took hold (Zuboff 2019). This cannot be blamed on the original versions of humanism, of course, not withstanding that they at least allowed in their conceptual apparatus for these developments.

In an all too short and schematic way I want to situate this first version of humanism against some modern conceptual and ideological emphases.

Humanism and Universalism

In my understanding old humanism showed universalistic ambitions: looking at this tradition as an anthropologist I identify universalistic aims that were borrowed from Christianity. That is to say, according to this religion, God created the earth and everything on it, including man and woman. In Genesis one reads that these first humans disobeyed their Creator and were therefore chased from paradise. They stand for the whole of humanity in this myth. When subsequently the Messiah (Christ as the son of God, in the Christian lore) materializes as a human being to sacrifice himself and thus save the creation, the new goals are made explicit: all human beings are said to live in the sinful position of Adam and Eve and will only be saved from eternal doom through their conversion to Christianity (with baptism, obeying the rules of the gospel, etc.). From that point on it is clear that Christianity has universalistic ambitions: the new convictions and rules hold for *every human being*. God did not just create the first humans and their kinship line, but all of humanity. In order to be saved, then, each and every human being has to be converted to the Christian reading of the Creation lore. This resulted in the obligation of Christian clergy to convert people worldwide, which is a coercive universalist rule.

In the words of the conservative philosopher Carl Schmitt, one speaks of the 'first nomos': the world and human beings are defined first and foremost in the words of the Jewish–Christian God. When later on (especially with the great explorers and colonialism) other and even unexpected branches of humanity were 'discovered' by Christopher Columbus, Vasco da Gama or James Cook a 'second nomos' emerges, in Schmitt's words. Obviously different traditions were detected, and they were looked at and described in the European concepts of the time. That is to say, they were thought of as existing in the pre-Christian era. The postcolonial thinker

Walter Mignolo (2015) discusses this short history sharply, concluding that the original humanistic view (in the sixteenth century) of universalism, partly unknowingly and unwillingly, supported colonialism and capitalist supremacy throughout the subsequent centuries by defending the presumed superiority (in terms of progress, development and so on) of the culture of the western conqueror. In doing so, the colonizer became the 'civilizer'. All other definitions of humans and societies were seen and described almost without argument as remnants of a 'former' stage of humanity. Indeed, human or cultural time was seen as a unique and linear progression first and foremost (the spatialization of time in the notion of the 'arrow of time'), as represented by the European nomos. The history of European culture henceforth became that of the predecessor of all of humanity; the arrow of time became the universal model for the history of mankind in general. Mignolo remarks in his critical postcolonial analysis: (but) 'the first nomoi were many' (Ibid.). It is precisely this perspective that was not picked up in the political–economic history that ensued: according to this ideology there is only one reality, and that is laid down in the Creation story of Jewish–Christian lore. Within that view what appears as different can only be seen as primitive or not-yet-developed. Such an interpretation gave Christians (and capitalists) the task to 'help evolve' those who were seen as lingering in a former era of humanity, which occasioned and excused the converting zeal and colonial exploitation throughout the last five centuries.

That was the historical context in which humanistic universalism emerged: humanists adopted this view without genuine (self-)criticism. The question is: does this universalism hold a flaw of reasoning? Personally, I think it does.

First of all, when looking at the history of universalism one lands in the Middle Ages: the then-fashionable theological–philosophical tradition of 'universalia' contained statements that held true everywhere, forever and under any condition, since they were thought of as *universalia a priori*. When interpreting the words of the Creator in the holy texts, what was found as truths had to be understood in this way: the 'first nomos' (the first norm, order) cannot be anything else but eternal and context-free. What differs from or cannot be integrated into the truths of the text should be disqualified as primitive, i.e. situated in the pre-Christian era of humanity.

A quite different interpretation can be defended, though. When we look at Michel de Montaigne (in the 2004 edition, paragraphs 31 and 49) we can see a genuinely free thinker in action. Montaigne witnessed how the first Indigenous people from the Americas were shipped to Europe to be shown at royal courts as curiosities. In contrast to most of his contemporaries Montaigne showed interest in these people, stating that their tastes and ideas were different from but not necessarily inferior to those of their European counterparts. He lived in Europe at a time of harsh religious conflicts and wars, and remarked in the paragraphs cited that the way one Christian (e.g. a Catholic) treated another one (e.g. a Protestant) could not be called superior to the ways these non-European human beings dealt with each other. He even saw more civilized views with them. But Montaigne, for all his humanistic intentions, was an absolute exception with this deeply respectful attitude and open-mindedness.

In a similar line of thinking one can search for universal features, over and across cultural diversity, by comparing in depth the norms, values, opinions and cognitive tools of different communities. This could then eventually yield *universalia a posteriori*, meaning that after painstaking empirical analysis and comparison one would conclude on what is common and what is particular in the many ways of humanity. The obviously open and respectful interest of Montaigne could show the way. However noble and open-minded this may sound, it is far from actual practice in religious and in most humanistic circles even today. Even when, after the Second World War, the concept of 'civilized Europe' against 'primitive others' was dropped from official political thought (and understandably so after the horrendous things the so-called civilized peoples did in that war), the world community decided to launch some UN daughter organizations with a quite similar old 'first nomos' mentality. When UNESCO (the United Nations Educational, Scientific, and Cultural Organization) and UNDP (the UN Development Programme) were launched the politicians in charge started speaking about the 'developed world' (meaning western, European-based communities) and the 'un(der)developed' parts of the world. The mission now became to 'develop' (literally to have them grown out of their childhood envelopes or baby suits) these others, so they could enter the phase of humanity which is seen as later in the uniform progression of time. They could be educated in the culture of European/western stock. Again, my point is not that the latter has no qualities; the critique is that 'others' were unilaterally and systematically framed as 'primitives'. No developed human could learn anything worthwhile from them, it should be understood.

At a time when the cruelties and the denigrating treatment of peoples during colonization around the world is more and more documented, the proud claim to superiority of western cultural values is in jeopardy (see, e.g., Ghosh, 2022). The fact that other cultural traditions committed similar crimes against humanity (as we call these feats lately) is not an excuse for one's own misdeeds, of course. But my focus here is on the self-declared superiority of western culture, including in its humanistic version. A short list of 'our own' mischiefs is enough to become conscious of the range of them:

- The well-known history of the Holocaust by a highly developed part of Europe, with active involvement (Nazi Germany) and passive collaboration by many countries, accounts for some 6 million victims.
- The colonial regime in Congo is believed to have cost the lives of 5 million.
- The Dutch regime and the colonial war in Indonesia are said to have cost as many lives there.
- Recently, on the occasion of the 500-year celebration of the civilizational campaign of Europe in the Americas (symbolized by Columbus landing there in 1492), several Latin American scholars broke the silence of five centuries: they reckoned that this incessant so-called Christian development programme had cost the lives of between 74 and 94 million Indigenous people (through manslaughter, slavery, famine, the deliberate spreading of fatal illnesses and so on: Fisher 2017). I will come back on this point in later chapters.
- The almost incessant wars in the European context (at least since the Roman occupation ended) resulted in millions of casualties and endless glorification of aggression. The slogan that 'man is a wolf to every other man' by nature (ascribed to political philosopher Thomas Hobbes) offers a striking metaphor in my mind.

I will rest my case here. Suffice it to say that the above list of genocidal incidents can hardly be seen as proof of a culture's moral superiority. Of course, one cannot blame humanists for such a history. What one should do, I claim, is to point to a big problem: the well-meant change in worldview of the old humanists was too one-sided in taking European/western superiority for granted and not being more critical about it. The consequence was that the deeply irrational racist and rogue self-image on which the said superiority was based was not made explicit, nor critically assessed by humanists over the past five centuries (with very few notable exceptions, of course). My understanding is that this did not happen precisely because a lot of the exclusivist mentality (and ontology) of the religious zealots in the humanists' context was not systematically and critically scrutinized. Nor did humanists question the universality they were dreaming of: they went along in forcefully universalizing what was and is in fact a particular, local set of values and insights in humanity and society. However, by omitting to criticize the universalism one justifies many of its effects.

Humanism and Cosmopolitanism

In recent years a growing populist discussion on cosmopolitanism has been emerging. According to the official doctrine of western foreign politics, such as the Hoover doctrine in the USA (Kissinger 1994), the international involvement of western democratic countries – called cosmopolitanism – aims at bringing democracy where it is not yet established. Some would grant that, since the Second World War, the worldwide deployment of multinational corporations has at the very least also been beneficial for the target countries, apart from spreading a generalized regime of freedom (Weitz 2019). Populists of rightist and leftist vintage have claimed meanwhile that cosmopolitanism in fact came down to forcing western values upon peoples elsewhere in order to protect western interests along the way. This undoubtedly overlooks the good intentions and hopes of many who participated in the process, but the devastating effects of western 'extractivism' (Ghosh 2022) cannot be denied here. In actual fact a particular format of collaboration and conflict resolution was mostly enforced on all in order to safeguard the interests of the few (see, e.g., Mattei and Nader 2014). What interests me most in the present book is how to overcome this 'blind spot' in the broad humanistic project of today: how to decolonize (western) humanism and Enlightenment thinking, and how to treat people around the world decently as fellow human beings sharing the earthly context?

The whole concept of cosmopolitanism is relatively new and could probably not have been anticipated by the old humanists (Bodelier 2021). If one looks at the practice of the last thirty years (the era of neoliberalism in the actual economic politics of western countries) it cannot be denied that power and money are being concentrated at a tremendous speed in the hands of very few people (Stiglitz 2011; Piketty 2014). Further on in this book I will draw on the evolution in the ICT (Information and Communications Technology) world where this quasi monopoly of a handful of players visibly leads to the destruction of democratic practices, much to the benefit of this small set of stakeholders. Equal rights through solidarity, and (economically) through redistribution of opportunities and wealth, are openly attacked or discarded as 'old socialism' by leaders (not only visible in topics by President Trump and the Republican Party, but also in the bitter struggle between so-called moderates and socialists within the Democratic Party in the USA, as well as in EU countries). This has led to the privatization of health services and large parts of education in some countries (e.g. the USA, the UK and the Netherlands), for example. In this type of politics cosmopolitanism is equated with 'economic globalization', with the fast development of a black market in the big cities (e.g. the so-called sweat shops, but also a large market in criminal drugs traffic).

At the same time, the financialization of significant parts of the economic activities in western countries threatens to unsettle the survival chances of large communities there and elsewhere (see Stiglitz 2019 after the banking crash of 2008–9). In that context humanism cannot avoid a critical political–economic analysis of the survival conditions of humanity today. This will inexorably lead to a refinement and a redirection of humanistic positions, where political–economic choices (as expressed in the discussion on cosmopolitanism) will have to be made clear. Today, some individuals are modestly exploring this expansion and redirection of humanism, but they are still very few in number (Maeckelbergh 2019).

Humanism and Liberal Democracy

Renaissance humanism did not have a clear and worked out idea of the political project that we now know as liberal or representative democracy. The latter is clearly a modernity project, developed gradually with and after Enlightenment. Humanists turned to the Ancient Greek texts stemming, as we know, especially from Athens and the first centuries of Roman government (basically the time of the Republic). The discovery of these texts allowed for thorough criticism of the autocratic and heteronomous model for society which had been installed in most regions of western Europe over the preceding centuries by the Roman Christian Church. This view on humanity fitted in with a stratification of society which was, after the fact, remarkably similar to what was dubbed the 'caste system' (when British rulers wanted to characterize the societal system in colonial India). That is to say, basically three social strata were distinguished. The top layer consisted of the nobility, which was overall a stable subcommunity that perpetuated itself through a system of privileges, marriage within inner circles and rights on the ownership and the produce of most land. The second stratum was that of the clergy: members of this subcommunity justified the continuity of social distinctions by their interpretation of the holy texts. Since they were schooled to a greater extent than the nobility and the third stratum, they in fact incorporated ideological control in power relations vis-à-vis the others. The third stratum consisted of the overwhelming majority of the population: they were the have-nots of the first fifteen centuries AD. They owned no land, were born poor and had to work the land of the nobility in order to feed their kin. They were fully at the mercy of the local noble families, who owned the land and everything on it (including the people). Individuals could very rarely 'move up'; in exceptional cases a helper could be freed and even given a 'title' because he had shown exceptional courage in the service of his noble lord. Apart from that, higher clergy and members of the larger group of abbots received profitable domains through kings and higher nobility, on the condition of controlling the local dependent subjects. Stories abound on how this practice of distributing profitable abbeys and suchlike was increasingly used to secure an easy life for bastard sons of the upper layer. So, all in all, society was to a very large extent fixed and closed: the lower part worked in order to survive on the land owned by the nobility and had to give part of their crops to the latter. Moreover, they had to provide free services to them too. In return they were allowed to live on the land and occasionally would get some protection against external enemies. The clergy provided the ideological frame which justified the relationships of dependency: on the authority of the higher power of the Creator the social stratification was explained (hence heteronomy). Only when cities began to emerge (simultaneously with house industry) in the north of Italy and in the Low Countries did the idea of a free citizenry emerge: citizens worked in shops they owned and sold the products of their labour. The first consciousness of freedom and, with it, the first more organized resistance against the 'caste system' – European style (my words) – grew in these cities. An important group of scholars in medieval cities like Bruges, Florence and so on produced tremendous evidence of this shift, which gradually allowed for the humanist thinkers of the Renaissance to emerge (see, e.g., Dumolyn and Brown 2019). With this extremely short paragraph I sketch the context out of which humanism emerged. The sketch of the context, and of the powers that were, needs at the very least one more amendment: one could say that the static and superbly ordered world of the Christian era was the ideal of the clergy and the nobility of those centuries in Europe, but did not necessarily mirror reality.

The philosopher A.N. Whitehead was particularly interested in the sequence of political philosophies in this part of the world and in understanding what could have been the real impact of them. I relate some of his analysis: Christianity in a sense attacked the scope of what was seen as human by the Ancients; it introduced the notion of a god who loved humanity as a whole and every human being in particular. This entailed the problematizing of the serfdom of Greeks and Romans, at least in principle. But the dominance of a despotic god-figure with an indubitable view on good and bad in human lives installed a 'slavish Universe' (Whitehead 1961: 29): in the holy texts the definite context of humanity was given, and it attached little value to freedom for everyone. The model that was installed in practice became that of 'coordination': 'The Church coordinated religious speculations; the Feudal system coordinated the intimate structure of society; the Empire -or, was it the Church?- coordinated the Governments of the provincial regions.' (Ibid.: 30). There is no room for any genuine freedom in such a model: intellectual activity was reduced

20 • Humanism Revisited

primarily to niggling wars of interpretation of unchangeable divine texts on harmless or next-to-irrelevant issues (such as the famous question: how many angels can sit on the head of a pin?). The coordination worked best in rural areas, and it weakened in urban trade circles. It was never free from being endangered at the empire level, where Rome was the continuity factor, but had to lean on lay power of all sorts along the way. For example, lacking a decent alternative, the Pope had to make Charles the Great emperor in 800 AD; a powerful northern 'barbarian' who did not even know Latin at any decent level thus became a major ally of Rome. When particular urban areas in Italy and Flanders grew richer and more powerful than the medieval control (sitting in the Vatican) could manage, humanist thinkers started to emerge there. Whitehead observes that the growth of humanism coincided with (or maybe reinforced) the shift towards a competitive view on human beings, i.e. first and foremost individual (male) citizens.

At the higher levels of social and political organization, the philosopher remarks, competition did not produce solidarity or freedom-in-togetherness (as some interpretation of humanism might have us expect), but forms of autocratic leadership. Examples abound: such bloody kings as Henry VIII or the rather paranoid and religion-driven Charles V, who raged throughout his life against freedom movements. The latter's dynasty and that of his son, Philip II, not only destroyed the growing wealth of cities in the empire (like Bruges, Ghent or Antwerp in Flanders) but also started merciless wars against any form of critique on Roman corruption, and thus inadvertently reinforced the political organization of Protestant–humanist circles. So, yes, humanism led to more systematic criticism of the coordination perspective which was incorporated in Roman Church rule, but it did not lead to social and political freedom. In the subsequent period of industrialization (starting in seventeenth-century England and expanding throughout the West in the period to come) the competition view on reality (and humanity in particular) did not work well, Whitehead remarks. Instead of guaranteeing freedom it generated a continuous struggle between privileges for an elite and the welfare of the total community. Thinkers such as Jeremy Bentham and Auguste Comte first, and Karl Marx and liberals afterwards, were inspired in different ways by humanist thinkers, but historically this led to terrible and continuing poverty for large segments of the European population most of the time: one thinks of the living conditions of the new proletariat right up to the Second World War. Here it is good to remind the reader of the recent thorough researches of Thomas Piketty: except for a short period after that war in the West (the *heureuse trentième* 1950–80, Piketty 2014) inequality was the rule rather than the exception.

Moving on to our era it is important to isolate one more concept that underwent a slow transformation through humanism: the value of *caritas* (charity). In the Christian tradition charity was a value, which was intrinsic to the religion of love, to be sure, but one that also stemmed from the heteronomous sphere: Christ reformulated the existing basic social rule to become first and foremost that of *caritas*. Christianity distinguishes itself from both Judaism and from the Classic Roman mentality along this line. All of humanity should be saved, according to the new gospel, which would show how deeply relevant 'love your neighbour', or *caritas*, is for the Christian belief. In his treatment of liberalism, however, Edmund Fawcett (2015) stresses how a humanist reinterpretation of *caritas* resulted in a double principle of tolerance and of solidarity. The former could be called a passive and the latter an active version of *caritas*. The former was to become central to the liberalism ideology, while the latter would be stressed especially by socialists, Fawcett argues. To this day, this bifurcation of the humanistic heritage seems to be dividing those who have humanistic views: though they share the non-theological value of caring about kinfolk, liberal democrats and socialists will be divided in the way they support the first or the second principle. In his synthetic analysis of humanism Kruithof (2001) stresses how this reinterpretation of a basically Christian value prefigures a principled distance from Christian heritage. That is to say, 'socialists' and 'liberals' have elements of humanism in common provided they focus on their shared critique of Christian attitudes and drop the ideological emphases that divide them (e.g. the solidarity value).

Looking back, I think one can say that hesitant initial steps towards democratic views and practices can be recognized in the old humanists. A heavy emphasis on freedom of thinking, speaking and choosing, and the right to determine for oneself what choices to make in life (with freedom of movement, but also freedom of religion), can be found in the first humanists (Francesco Petrarca, Erasmus, etc.). This entailed a series of long and often tedious struggles to free oneself from Church, gentry and even family ties, at a time when the first signs of relative prosperity became visible in some European cities (Frankopan 2023). A long battle against repression ensued, with witchcraft and heresy accusations and with often devastating wars and punishing raids (e.g. the fierce campaign of the Habsburg regime in the Low Countries in the sixteenth century, brutally ending the international role of the city of Antwerp; Olyslaegers 2020). These power conflicts kept growing, at the least through the so-called religious wars, for about two centuries – and some even through a good part of the twentieth century, when secularization became mainstream in most countries of today's European Union (EU) (but not all: see Poland, Hungary or Greece, for instance). Along the way many humanists had to fear for their life: to mention only the 'figure-heads', Giordano Bruno was burned at the stake because of his predilection for free thinking; Galileo Galilei is remembered also because he was forced to choose against scientific truth in order to save his life. It was just a few generations further down the road that some of these ideas and attitudes broke through in political thought, especially during the Enlightenment period with the American and the French Revolutions.

The liberal political programme, which was developed from the eighteenth century on, was concretized in experiments with representative democracy. For generations the right to vote and the right to participate in elections as a candidate was reserved for men, who had property (first and foremost landowners). Piketty (2019) gives a detailed analysis of the way this prerequisite of ownership (land and, later, other types of capital) has been conditional in all the democratic states from the beginning, and often works through as a barrier in the costs of campaigning right up until today. Neither the radicals in the American or the French Revolutions nor their 'softer' political colleagues ever understood democracy in a stricter sense of power to all, regardless of economic background or position. So, when we speak about the notion of freedom, set forth by humanists, we must be careful not to overstretch the content of the term: private property was and is still rather generally and very generously protected by political representatives, which has set important limitations on freedom practices ever since humanists (re)started referring to this value (see de Dijn 2020).

Thus, in the argumentation of many liberal thinkers right up to the present time, the new territorial reasoning in terms of nation-states is a consistent expansion of the old (medieval) thinking about land and rights: for example, Eric Weitz (2019) states that even in an era of internationalization and globalization the constitutional nation-state is the obvious and necessary guardian of rights in order to safeguard national, but also international, continuity. In that light the fierce conflicts we have witnessed since decolonization (right up to the seemingly endless wars in the Middle East, Afghanistan, etc.) are only 'birth pains' of nation-states in formation, according to Weitz. When one sides with these ideas (and Weitz very explicitly calls himself a liberal democrat), the notion of humanist freedom is historically a source of inspiration.

Looking at the present era I think it is not farfetched to point to two recent developments that seem to 'empty out' even more the old humanist notion of freedom and free choice:

1. the fairly recent but rather rapid growth of extreme-right movements in all parts of Europe and in the USA (to limit myself to just these areas of long-term democracy and humanism) expresses deep mistrust in humanist–democratic views on humans and society. The coupling of this political rejection of democracy with a renewed fundamentalism of religious vintage is certainly new, even if the depressing atmosphere of

the COVID-19 pandemic (lasting over two years) may add substantial power to this drifting together of desperate people. The theological department of Humboldt University in Berlin, Germany recognized the coupling of extreme-right ideas and religion as problematic and organized a symposium to gain an overview of what is happening throughout Europe (and, to a certain extent, the USA) along these lines; the result of this overview is rather stunning (Höhne and Meireis 2021). The coupling between fundamentalist religious organizations and extreme-right political movements is widespread on the one hand. Some forces serve as go-betweens in order to destabilize European democracies and to enhance the said coupling. In several cases it is documented that 'weaponizing' and the organization of paramilitary training camps in a few EU members states has been going on for a few years (Höhne and Meireis 2021; Ponsaers 2021). The dark net of the major internet providers is an especially forceful and rather efficient means for mobilization, it seems. The growing power of extreme-right political parties who participate in elections, from the USA over the EU (e.g. Hungary, Poland, the Netherlands and Flanders-Belgium) to economically important countries like Russia, Brazil or the Philippines, allows anti-democratic groups to gain power without many restrictions.

Apart from this clearly anti-humanist and anti-democratic movement, more or less official or previously respectful fundamentalist groups launch rather effective actions against secular society. A few examples illustrate what is happening in this field. One of the fundamentalist groups is Ordo Iuris, originating from a Catholic South American group of highly trained juridical specialists. They studied the constitutional system of Poland and subsequently successfully started to attack any claims from secularists, but also from the LGTBQ+ movement, in order to turn around Polish laws on the basis of juridical arguments: these laws would be in conflict with the Polish Constitution, according to Ordo Iuris. Hence, on the basis of juridical scheming the group managed to bypass democratic parliamentary changes in the laws and 'restored' a so-called constitutional ban on progressive law making.

A similar, but purely political–ideological attack by extreme-right groups in Germany and France (Generation Identität and Génération Identité, respectively) focuses on governments and representatives of democratic and humanistically inspired representatives in different EU countries. They claim that the latter would violate the 'cultural identity' of their country. Thus, political representative would subscribe to the idea of *Umvolkung* (a Nazi concept which claims that local people will be outnumbered soon by 'others' coming from elsewhere). Again, these movements claim to be defenders of the deep Christian tradition of this part of the world, thus linking their anti-democratic programmes to an anti-secular view on man and community. So far, governments do very little to fight these movements or to call on the responsibility of the internet providers, even if the latter allow them the use of the dark web to spread their ideology.

2. The second trend which weighs on the nation-state as concretization of at least one way of structuring the humanistic–democratic perspective is the globalization of the past few decades. Although the free-market view originated in the early phases of the nation-states, especially since nineteenth-century industrialization, this way of thinking and acting in the world has grown rapidly and rather aggressively in the past eighty years or so. It then transcended the nation-state with the birth of so-called multinational corporations, promoting what is now known as the globalization of economic activity: both types of resource (raw material and labour alike), trade and business relations, and actual markets are acting on the planet as a whole. This applies to the capitalist expansion of activities as well as to the so-called communist international deployment of economic activity (e.g. the 'new Silk Road' of China: Frankopan 2018). This economic globalization established what I called earlier worldwide interdependence in certain, often vital, sectors: humanity is gradually moving towards the dependence of all on all in order to survive. The mere blocking of the Suez Canal by a large container ship in 2021 illustrated this point, causing disruption in a series of countries because of the unavailability of parts that were held up for weeks. The so-called 'chips' crisis after the lockdown of 2021 was another illustration: relaunching economic activities in several developed countries was retarded because the production of microchips in two or three 'cheap labour countries' could not cope with the demand. Thus, factual interdependence in economic processes became clear to everyone, but the ideological switch to recognize this change in perspective (in comparison with the nation-state view) still lags behind.

In a general way this rapid economic globalization produced new poverty groups (which can only survive by earning a living through cheap labour) in many cases, and a new structural inequality between haves and have-nots all over the world (Piketty 2014). Groups of 'rejected' people (Sassen 2016) seem to look for political support from the new populists, who started attacking cosmopolitanism as the ideology of the rich. In practice, the digestion of economic globalization seems to have thus far triggered two different reactions at the political–ideological level: on the one hand a group of highly trained people move around in a separate and often virtual world of opportunities, and see the benefits of cosmopolitanism based on economic privileges; on the other hand a quite separate set of groups, with modest education levels and low incomes, turn away from this world and are tempted to listen to the old sirens of nationalism. The political nomenclature in the West seems to be split along the same lines: in the USA, the UK and all countries on the European continent a large minority, and sometimes a small majority, of voters tilts towards extreme-right or otherwise neo-nationalist programmes, which has made each and every formation of a government for the past couple of years (openly or silently through compromise) something very hazardous. For example, deeply democratic countries like Belgium, the Netherlands, Germany or the UK have seen growing ideological strongholds which openly defend the earlier-mentioned theory of *Umvolkung* (Höhne and Meireis 2021, Pinxten 2021, Ponsaers 2021).

On top of all that the so-called neoliberal ideology within some of the mainstream political parties and organizations (such as the EU; the UK; and, of course, the USA) attacks the primacy of the political over the economic sphere. Starting with the Reagan–Thatcher era it successfully diminished or even destroyed forms of redistribution, since that would make people lazy and hinder the expansive activities of the strong economic agents whose wealth is then said to eventually benefit all (by the so-called 'trickle-down' effect). The often-heard slogan that 'greed is good' then comes to the fore. Of course, at least some economists have proved that this last promise is false since a new, structural inequality is growing fast over the past decades under the aegis of global neoliberalism (Piketty 2014 and 2019, Stiglitz 2019).

I mention these contextual changes in the economy and in the mentality of people in order to discuss what is happening in the West, as the region where the humanist view on humanity and the world emerged precisely as a critical reaction to the domineering of people by one ideology (i.e. religion) and the political–economic elite using it to subordinate a majority for ages. In this changed context in which we are living today, one of economic globalization, criticism of this new format by humanists is lacking. My inkling is that either the turn to the dominance of economic thinking (the *homo economicus* concept) is shied away from because this would lead to a split between liberal and socialist inclinations or else humanists do not understand the deep changes this turn brings about, meaning that the broad views on humanity and on freedom of choice are practically erased in this new context. If this suspicion holds, then it is high time that the critical attitude which is intrinsic to the humanist tradition was awakened again.

In order to substantiate these statements, I need to offer more data on the changes I am exposing here. Recently, several studies on the waning attraction of democracy (and hence of humanism as a broader background) were published. The well-known Pew studies in the USA quantify this trend (Pally 2020), which for me urges concerned humanists and other democrats to look closely at the messages we are sending out to a larger public. Political studies try to analyse what has been happening in recent years and why the message of the past few centuries seems to lose strength. Some of them are the talk of the town ... in academic circles, like Steven Levitsky and David Ziblatt (2018), and influential retired politicians try to scrutinize the present situation (e.g. Albright 2018). Still, like the works cited, a lot of these tend to be ideological messages in the first place.

Apart from this, strictly scientific research is growing. Some anthropological studies try to understand how and why the democratic project remains a difficult one. For instance the book *The State We're In. Reflecting on Democracy's Troubles* – edited by Joanna Cook, Nicholas Long and Henrietta Moore (2019a) – offers deep insights. The editors make clear from the beginning that the concept of democracy was never clear and straightforward for all involved. Socialists emphasized the force of a correct and just nature of redistribution of wealth, with the state in an important role of powerful go-between. Anarchists partly agree, but they will argue that the role of the state should be downsized continuously, since concentration of power in such a structure will inevitably produce corruption and abuse of power. In the present time the growing criticism of representative democracy (with elected parliaments and suchlike) apparently allows for a moderate rebirth of anarchistic analyses: Occupy, Indignados, new commons, Extinction Rebellion and so on point in that direction. With the hardly different and rather spontaneous forceful actions of climate alert groups, this trend might even become directly influential on the political scene. Margaret Maeckelbergh (2019) captures the trend by saying that a growing group of people and organizations in primarily democratic countries started experimenting with 'horizontal' political structures. Sometimes apart from, sometimes within and on still other occasions in opposition to existing 'vertical' political structures (top-down, as is the case in practice with most representative democratic systems), the 'horizontalists' start commons, neighbourhood networks of self-regulation, urban and interurban platforms of decision making and so on. Some would call these initiatives post-democratic, since they take power into their own hands and want to strive for a good life for all independently of elected representatives. However, the open interactive form within these initiatives, the limits in time and space of any mandate in such horizontal structures, the freedom of speech and writing, and so on guarantee a very basic understanding and practice of democracy. So, it might be more correct to keep talking about experimenting 'in' and not 'after' or 'against' democracy.

Especially in rich countries the intrinsic risks of globalized economy, such as were illustrated in the financial crisis of 2008–9, gave a boost to alternative economic initiatives and direct democracy. Almost overnight, urban areas, especially, saw the booming of economic alternatives such as the doughnut economy (Raworth 2017), new commons (Bauwens 2013), the circular economy, the city as most powerful engine of community formation (Corijn 2017) and so on. A long-term researcher in the field of democracy and authoritarianism, John Borneman (2019), launches an appeal to all democrats to at the very least amend the old top-down democracy (with sparse, relatively indirect corrections through the election systems) by means of small-scale, social, scientific studies that would sustain a more bottom-up and relatively permanent impact on governing practices and policies. In the second part of this book other suggestions will be discussed.

Chapter 2

WORLDWIDE INTERDEPENDENCE IN THE TWENTY-FIRST CENTURY

It is not popular to warn against crises that are emerging, or that we are stuck in already. The author risks being dismissed as one of those alarmists or doomsday thinkers who threaten to spoil the party. In the past few months I, together with several of my critical friends, were also classified as 'woke' by rightist politicians. That seems to have become an argument in itself now: when you are critical about western values and actions, you are woke in the sense of 'the enemy from within', apparently. However, as a social scientist it is my inclination (and maybe even my duty) to situate values, perspectives, facts, models and theories in the contexts in which they exist. This implies that I must critically assess the contexts and the changes therein, in order to give a valid interpretation of the phenomenon I want to look at (i.e. humanism today). And the present context, as I see it, is subject to deep changes: Europe and the West are no longer the hegemonic powers they used to be in the centuries after Columbus. Therefore, many parts of the world do not as readily look up to them as the unique example to imitate in order to reach a better or higher level of humanity (as the colonizers pretended). At the same time, some unwanted negative effects of the western exploitation of other communities and of the earth have become clear today; elsewhere (Pinxten 2019), I have listed at least five major crises that seem to shake the worldview we have been cherishing for centuries. Climate warming, destruction of biodiversity and new structural inequalities may, by now, be the better-known derailments.

Of course, a social scientist may be considered to have insufficient competence to really make this sort of statement. It is good, therefore, to refer to powerful think tanks where such expertise cannot be denied. The group of intelligence agencies of the United States of America, known as the National Intelligence Council (NIC), publishes every five years a synthetic report on the state of the world. Only recently this body brought out a prognosis for the next twenty years: the *Global Trends 2040* (NIC 2021). This is an intriguing and powerful source of information on the main issue of this chapter, even more so since the report carefully refrains from broadcasting ideological statements. For example, I could not find such terms as 'free market' or 'enemy' in the document. On the other hand, this official conglomerate of agencies seems so convinced of the gravity of the state that the world is in that it organized a deep, interdisciplinary collaboration between each partner to reach a substantiated view on the subject matter. Natural scientists, demographers, economists, ecological scientists, climatologists and political scientists pooled their respective competences. The subjects treated include all sorts of important themes: which natural resources will become crucial in the next decades, what developments in IT (information technology) can be expected, what is the foreseeable impact of ongoing climate change on the living conditions in different parts of the world, and what effects will these issues have on migrations and political reactions everywhere? The report ends with a synthetic proposal of five distinct scenarios for the next two decades. At one extreme there is a positive and collaborative network of ties throughout the whole world, showing awareness of the factual, deep interdependence in the present era. At the other end there is the horror scenario of a deeply divided world where smaller political units are in a state of permanent war with each other. Given the fact that this consortium of agencies has many times more means and personnel to guarantee a substantial overview of facts and models on such a vast subject than individual researchers can ever dream of, its analysis is greeted by me (and several colleagues I discussed this with) as a major argument in favour of the cautious conclusions I had reached through my modest personal search.

In a few paragraphs the NIC does express the wish that something like a 'free world' will be a main source of influence in the coming years, and will thus allow co-determination of decisions on the management of any of the crises mentioned. Personally, I would be rather more critical about the ethical stands of some of the big players in our world (especially the major financial groups but also the new giants in the ICT world, for instance). The lesser or greater role of public institutions in the control and maybe sanctioning of the 'giants' would then differ from the rather hesitant attitude of the NIC. Still, the report discussed here recognizes the illiberalism that is emerging in the so-called 'free world' and warns especially against the actions of neoliberal economic circles, who continue to eliminate redistribution mechanisms and institutions to the benefit of limitless privatization. My personal addition would be as follows: starting from such analyses, and the occasional syntheses by political economists (e.g. Piketty, Stiglitz) and social scientists, I can sketch a picture of the state of the world (the context) over several years now. The rest of this chapter comprises a short synthesis thereof.

As said, working as a contemporary social scientist I emphasize the relevance of context when describing a social/cultural/psychological phenomenon. That entails, for example, recognizing that the rather general belief in determinism in the nineteenth century (economic laws, but also evolutionary laws with a deterministic flavour) should be amended today: whatever else, I am convinced that physical and ecological, but also political–economic, environments certainly co-determine meaning, values and other cultural forms and contents. But they are not the sole factors. The belief in the time-independent causal laws of classical determinism should be understood in the context of the nineteenth century, and needs amending today (see, e.g., Prigogine and Stengers 1984). There seems to be a small margin for human creativity instead: the human species has at least some room for phantasy as well as for outright lies, one should admit. For one thing, humans create unobservable and supernatural beings and subsequently allow these creations of the mind to harass them or dictate how they should live. There is probably no other species we know of that does similar things (Pinxten 2010). Secondly, as I mentioned before, the human species has for at least 300,000 years now been spreading genetic material all over the planet (Relethford 2006). Exchanging, borrowing and recombining genetic material seems to be a fundamental characteristic of *sapiens*. Finally, cultural elements are exchanged, appropriated and integrated with time between communities that have contact, directly or even through intermediaries. A recent instance of that fact is the expansive, often raw, globalizing spread of capitalism – made stronger and more efficient through its use of scientific and technological knowledge in the past couple of centuries – introducing a particular ideal of humanity with the *homo economicus*. The latter version of *sapiens* has universalistic pretensions, and causes global problems that may threaten the existence of many life forms on earth, including that of the human species. This is all part of the context humanity shares more and more today, and yet it was mostly unknown to our ancestors right up to the Second World War. Empires grew and collapsed in different parts of the world; the emergence of one global empire is a possible new development, but it might collapse as well when the old rules and principles are applied blindly as if they were universal.

In terms of the scope of this book, at least the past three centuries have also been those of the rise of the model of liberal democracy in a significant part of the world, as part of the heritage of the humanists. At least parts of this model (the electoral procedures, the acceptance of human rights) have been a frame for, and have thus politically formatted, the global expansion I refer to. Therefore, I think the humanist (and the democrat) of today should be interested in a critical analysis and assessment of her own worldview, and should even feel the urgency of such self-search. Rather than seeing the present format of globalization as the next step of western civilization, looking for new perspectives may be on the agenda.

Today, in contradistinction with only five centuries ago, humanity is becoming factually interdependent. It is amazing to think of the time lapse: *Homo sapiens* has been around for some 300,000 years, and the cultural–economic regime of the systematic impactful exploitation of natural resources has been ours for about three centuries (beginning with the Industrial Revolution: Frankopan 2023). This means that within 0.1 per cent of the existence of the species it changed its context in such a fundamental way as to become interdependent. Of course, to name just a few positive effects, life expectancy rose substantially through the development and spread of modern medicine, the demographic expansion took place (and needs managing today) and the cultivation of food grew tremendously but confronts humanity ever more with a deep need for equal and fair distribution on a worldwide scale. Amazing technological inventions should be pointed out, but many of them led to material products that have no genuine welfare value and that pose problems for the environment to the point where the survival of other species is threatened. Just think about the strange culture of consumption: that has one buy new clothes twice a year, yielding tremendous waste dumps and the pollution of the oceans with 'plastic islands' along the way. Or think of the fact that rich countries throw away about one-third of all good, usable food year after year. A strangely short-sighted interpretation of freedom in the ideologies of the free market and (individual) freedom of choice of the industrial period in the West (de Dijn 2020) seems to lead inexorably to an unliveable world of greed, conflicts and climate disaster in this so-called post-industrial era. When I try to systematize 'problematic developments' in this sense, I isolate five major global problems (and their interrelations) which should urge us to rethink and reorganize now. I summarize.

The major reset that is called interdependence: since the Second World War the military and safety policies of the West (and of communist and other, decolonized areas in the world) seemed to be struggling with the integration of two extremes – national or cultural identity on the one hand, and global perspectives on the other hand. European countries, who fought each other fiercely in that war, merged into a cross-national entity (now the European Union – EU) to which they gradually dispatched more and more of their formerly 'national' competencies. After the war, NATO (the North Atlantic Treaty Organization) emerged, which subsumed a great deal of the national military power of its partners under a common, also nuclear, umbrella. A few decades later the Schengen agreement coordinated much of the police planning and action procedures at the European level. With the Lisbon agreement (1992) the EU member states agreed to have their national budgets and economic planning screened by European services first, prior to discussing and officializing them in the national parliaments. Meanwhile, the Euro became the currency of a majority of member states, making their financial fate dependent on European instead of national decision making. After the financial crisis of 2008–9, definite plans were developed at the level of world players such as the International Monetary Fund (IMF) and the European Central Bank, but also the Davos World Economic Forum and the Bilderberg Group, to curb local or even national freewheeling in the name of national free-market prerogatives (Stiglitz 2019). In a similar way the World Health Organization (WHO), the EU and other international players claimed and received transnational power in the pandemic crisis that started in 2019 and was still ravaging the world by the end of 2021. Finally, because of its urgency, countries from all over the planet gathered and made more or less binding agreements on rules and principles in order to 'manage' disastrous climate change (the COP conferences, G7, etc.).

To be sure, such international and eventually global agreements do not yet have a high priority in the minds and hearts of most citizens in many states around the world. But the consciousness that we are presently depending for survival upon each other at a worldwide scale is growing. The youth of the world, especially, have shown this consciousness of global interdependence clearly and forcefully in their continuing actions by organizations such as Youth for Climate. My understanding is that the factual interdependence is rather rapidly mirrored in the minds of the new generation and is triggering new ways of thinking and of economic planning. At the same time, and with the same feeling of urgency, one sees a great deal of nostalgic, poorly informed and hence conservative reactions. One of these lines of 'coping' can be seen in identity politics, crystalizing in extreme racist and neo-nationalist movements (Ponsaers 2021; Höhne and Meireis 2021).

At yet another level, the new branches of an economy heavily centralized through ICT claim free-market heritage and yield excessive concentrations of wealth and power in the hands of very few. They also fall back on systems of neoliberal labour exploitation that threaten to do away with all the social and long-term developments made over the last century (Zuboff 2019). If anything, this development shows a second conservative political trend: exclusive private-property rights and the dismantling of so-called social peace and redistribution agreements, creating greater uncertainty in the job market and more inequality.

Thus, two opposing political and cultural views seem to colour the future of interhuman and community relationships. There is no doubt in my mind that in the interdependence view humanism, as a driving ideological foundation in several western countries, is called upon to become an ally. At the same time, the old humanism is being fiercely attacked by right and extreme-right movements as a mistake of the past. Here, too, the heirs of the humanist tradition (and hence of the democratic adventure over the past couple of centuries) should wake up and do their homework. This short mention of currents in the socio-political landscape invites us to take stock of the shifts which are redesigning the context we live in today (as quite different from that of three to five hundred years ago).

Summing up, the new contextual background of factual interdependence we are rapidly drifting into is showing at the very least the following crises:

Climate Warming

After a generation of denial and even 'fake' scientific research to sow doubt in the minds of less-informed citizens, most knowledgeable people today accept the earnestness of the situation as laid down in the reports of the UN agency IPCC. A leading scientist of this conglomerate is Jean-Pascal van Yperseele, climate scientist at a Belgian university. He summarized many of the reports (www.ipcc.ch in van Yperseele, Libaert and Lamote 2018). If CO_2 emissions continue at the same volume over the next few years some absolute horror scenarios can be outlined: densely populated areas in the world (a delta region in India of ca. 300 million people, parts of Central Africa, etc.) will be uninhabitable because of continuous high temperatures (above 40°C), the dying off of crops due to lack of rain or soil water, huge famine periods, massive gulfs of refugees who seek to escape from their uninhabitable area, and flooding of densely populated urban areas (like New York and parts of Bangladesh – but also the Netherlands, to mention just a few examples). The measures taken so far by the COP conferences will turn out to be insufficient. China, the USA and the EU, as main economic powers, take decisions but fail to implement them with sufficient rigour or responsibility. Power prerogatives and/or economic interests in the short time block a deeply efficient response, it seems. On the positive side, one should mention bottom-up initiatives in many countries which courageously steer towards responsible and sustainable

34 • Humanism Revisited

economic choices (commons, doughnut economy and suchlike: Raworth 2017, Lotens 2019).

Ecological Destruction

The reduction of biodiversity in the world is enormous. Species of all sorts are starving or simply killed off in the continuing trends of land grabbing: felling large trees for wood and for grazing land (for multinational burger chains and other big players); polluting seas and oceans all over the planet, which turns them into 'dead seas' through the chemical destruction of all life over hundreds of miles from the local delta; forcing peasants off their land in order to turn it into huge monoculture plantations or cattle fields for the meat industry; and, of course, massive destruction of land through the mining of gold or fossil resources. Obviously, this terrible trend, yielding the elimination of species at high speed, is more complex than I can sketch in these few lines. It suffices to state the problem, not to detail it; a vast literature is available.

The New Structural Inequality

Structural inequality occasions another global crisis (Piketty 2014 and 2019). After the Second World War a short period of redistribution of wealth was reached in most western countries, as a result of massive social protest. This switch is captured by what Thomas Piketty calls 'l'heureuse trentième' (the happy thirty years: 1950–80, Piketty 2014). But from the 1980s onward neoliberalism came to power: US President Reagan and UK Prime Minister Thatcher were to become the political pathbreakers of this ideology. Inequality rises sharply from that period on, yielding a new economic elite. They were instigators of the financialization of large parts of economy, accompanied by massive tax evasion (and the start of 'tax havens'). So-called new values were promoted: attacks on solidarity, the promotion of extreme competition between individuals (from kindergarten on: Standaert 2021) and a focus on values such as 'greed is good'. This led to strange anti-humanistic projects, such as the one studied by a fellow anthropologist, Sea Steading, as the development of islands for the happy few in oceans far from any claims or influence of national policies (Simpson 2013). The literature on inequality is growing fast: after pioneering work by Anthony Atkinson, both the IMF and the Organisation for Economic Co-operation and Development (OECD) recently joined in warming that too much inequality is poisonous. Piketty's thorough overview studies draw the general picture.

The Exhaustion of Resources

Natural resources of all kinds are limited. Oil and coal, one readily knows, took ages to be transformed from the original organic material they once were. As far as oil is concerned, we, mostly western agents again, burned most of what was easily and cheaply dug up and transported in the course of only one century. Gold, silver, manganese, cobalt, uranium and many other natural riches are being won from on and deep under the soil, from the bottom of the oceans and from mountains that we rapidly blow to pieces with dynamite. The rubble, as well as the rest of the extraction, is piled in huge waste dumps together with so-called used stuff of all sorts that has been replaced quickly by the consumer objects our economy keeps on producing. Timber is cut because we need hard wooden closets to store away the new stuff we are urged to buy as much of as possible by the mar-keteer. Not participating in this race rapidly pushes one into the reject pile of old-fashioned and more or less marginal human beings, who have lost the 'touch with the real world' as commercials will convince us. What is rare is reduced to its economic value (gold, but also the wilderness of the safari business). Recently, air and water have been drawn into the market logic and privatized, as land has been for centuries (starting with the indus-trialization of England: Piketty 2019). Even the DNA material of faraway tribes is patented today by 'clever' pharma representatives: the reasoning is that it might someday prove to be an advantage to the 'owner' if and when a particular pandemic medicine might be based on it. Umberto Mattei and Laura Nader (2014), who discuss this ethically puzzling development in juridical terms, speak of generalized 'plunder'. Such interesting human behaviour, says the sober anthropologist in search of new exotic stories, is driven by short-term profit protected by laws on private property and by the belief that growth can be endless, meanwhile disregarding the fact that we are exhausting natural resources at a tremendous speed (Hickel 2021).

The Demographic Factor

This is a possible fifth source of crisis, provided we continue to misman-age demographic forces. In the 1960s the general cry was that abortion was murder, and even contraception was forbidden by many Christian and Islamic missions. According to some interpretations of the holy books, humans should not interfere with fertility except by reducing sexual activi-ties. Today we know that demography is not equal to genetic potentiality: epigenetic factors play a very important role. Put clearly, it is not the case that all women in the world will produce ten or more children when they see the opportunity. Keeping all of them alive (through better medical care) would then cause a tremendous demographic curve, the specialists of years back pretended. Hence, this would call for disasters like famine, joblessness and the like, the Malthusians among them would state. Recent research adds more parameters and hence finesses the picture. An overview paper in *The Lancet* (Vollmet et al. 2020) sums up a set of contemporary and rather substantial data: it is absolutely clear from large survey studies that demographic changes and expectations of life and welfare interact intensely with each other. Given solid survival contexts (like retirement pensions, steady work, etc.) the demographic curve flattens quasi automatically.

Thus, without any serious policies for redistribution of wealth the world population is estimated to grow to 9.7 billion by 2064, flattening to 8.8 billion by 2100. However, when actively promoting child care, social welfare, policies of contraception and prolonged education of girls throughout the world, the curve will flatten considerably more, ending at around 6.6 billion by the end of this century. Thus, it is important to look at and secure policies on the combined factors of fertility control (through contraception, etc.) and the redistribution of wealth (Vollmet et al. 2020). This is a directly relevant point of concern for humanist activists: not only freedom of (individual) choice or the availability of medical care matter, but also the socially responsible rulings and adjacent institutions that accompany them.

Conclusion

Having listed five global crisis domains, I have sketched in a very rudimentary way some deep changes in the context in which humanism–democracy is operating today. It will be clear to the reader that this context has changed dramatically from the one in which humanism and its later offspring of Enlightenment came to life – especially since this worldview and policies, guided to a considerable extent by humanist thinking after the decline of monolithic Church power (starting in the fifteenth–sixteenth centuries), were to a large degree motivators of the western conquest (some say 'colonization') of the whole world during the three past centuries. If not at the heart of motivation (and indeed the first seafarers appear as zealous missionaries, with Columbus as the icon), the humanist European mentality was used to justify the conquering of the world. Or at the very least it did not fight it, let alone prevent it. It is necessary, in view of the damage we have now become aware of, to reset the guidelines substantially and make them appropriate to what we learn about the context of today.

PART II

A Plea for NEED Humanism

Chapter 3

NEED HUMANISM

Generally speaking, the primary school I attended was a decent school. It was situated near the popular neighbourhood I have already referred to and drew children from the lower social strata primarily. In my class there was one son of a physician. He was the only child with an upper-middle-class background. Very few of the boys went through any form of higher education (high school and on); only one (me) studied at university. The school was located in the so-called nineteenth-century neighbourhood, which became so typical of all the industrialized cities of the past century in Europe. Slums, cheap housing, street children, alcoholism and drugs (more so in the late twentieth century) would be typical of such first-generation industrialization, then and now (Davis 2006).

I described the neighbourhood in a previous chapter, pointing to the ill fate of one out of five of my classmates in elementary school. These little histories offered a sketch of what it meant to have grown up in such a neighbourhood of a medium-sized city during that period in Europe. The attempts by local and national politicians to elevate or emancipate the 'problematic youth' of lower social groups by means of education led to several initiatives of what was then called 'the democratization of education'. Child care and youth care emerged on the side. Of course, democratization of schooling was considered a necessity at that time: after many generations of neglect and minimal education of the people living in these quarters, the industrialization of the Antwerp harbour area became a priority after the war and hence more brains and better-trained workers were needed. The 'material' at hand, however, was the kind of low-class children I have described. My brothers, who went to the same primary school, had similar experiences with their classmates. As far as I know, the

40 • Humanism Revisited

neighbourhood I sketched was mainstream for any city at that time. Today, the same neighbourhoods house a new underclass: the immigrants and refugees took over where a high percentage of the former 'White' group had moved out to suburbia. In my professional career at Ghent University I was the head for some twenty years of a research and training group on 'intercultural education'. Our mission was to try and promote integration through better education of these 'non-local' new citizens and their children, living primarily in the same nineteenth-century poor neighbourhoods. Dropout and failure rates are very similar too, now as then. When looking at relevant literature on other cities in Europe, the similarities are striking (see, e.g., Bourdieu 2004). In that light it is, obviously, tragic that the 'old poor' can be led to believe that the 'new poor' are the reason for all the uncertainties in life.

Indeed, together with the new growth in socio-economic inequality, an ideology of 'cultural identity' started gaining power under the aegis of anti-democratic and anti-humanist ideologues (strengthened by Samuel Huntington's book calling for a 'clash of civilizations': 1996). My interpretation of the rise of (extreme-)right identity politics draws on my frame of reference as an anthropologist: the context in which ideological choices and value discussions are framed is progressively stressing a choice for vaguely emotional, alarming political 'solutions' and away from a broader and more rationalistic, humanistic optimism. I have emphasized the similarities in my own background context and those of the present immigrant and refugee population in our cities because both are so striking to anyone who wants to take contexts into consideration: it was not the sin or incapability of the 'old poor' that kept them out of the society of the wealthy of that time, but rather the privileges of the rich. An honest political analysis today should learn from the mistakes in that past period to develop good and effective integration policies today. Instead, cultural-identity ideologists 'blame it on the victim': according to them, it is the culture and morality of the latter which is identified as the cause of misery, and hence of the threat to wellbeing of the somewhat better situated today. The victim does not have the same cultural background and is therefore incapable of contributing to the 'progress' that the right cultural tradition stands for, according to this view.

In the light of the changing context I sketched by means of the growing crises in the contemporary world, I invite humanistic thinkers to critically reset their old principles. In my understanding, an honest self-critical update of humanism will help to address the tremendous problems we have before us, rather than look away from real causes and continue to blame it on those left behind. Within anthropology at least half a century of research has taught us that cultural identities are fluid, shifting, constantly reshaped and adapted through intercultural contacts of all sorts. In the light of this knowledge, any reference to ironclad, essentialist 'cultural identities' can only be ideological, inviting people once more to think along the lines of the us–them opposition. In view of the changing context humanity is drifting into, such a move leads to disaster, extreme conflicts and what humanists of any era will abhor. In the light of this I launch the positive or constructive call for a NEED humanism.

Humanism and the Ontological Critique: Modernism or Something Else Still?

The step in my reasoning I am about to take now may come as a surprise to some. I propose that humanists in this neoliberal and dangerous era should critically analyse their own ontological roots. Far from an endless journey in a kind of philosophy that delves into textual details (as a Kantian, a Hegelian or whatever else), my invitation is to look at the intuitive or ontological basis of the western worldview. Looking at the western tradition as an anthropologist, I observe that the ontology of traditional or original humanism was adopted basically from the religious worldview which went relatively unchallenged through modernity and colonialism, right up to our present era. That worldview and its ontology was firmly established in the learned world of western Europe back then (until the fifteenth–sixteenth centuries) for more than a millennium. The first humanists were Christians (and, in a related and slightly different sense of 'humanistic', also Islamic thinkers in preceding centuries – up to the eleventh century, according to Arkoun 1982) and their worldview was, and remained, deeply entrenched in this local, Mediterranean worldview. They were ontological dualists who took for granted that the world of God and that of humans had fundamentally different characteristics. And they were dualists in their ontology about the netherworld: what was called human was substantially, in a principal way, different from what was seen as 'nature', broadly speaking in the frame of reference of Genesis. There is a particular relationship between humans and the rest of nature in that explanation based on how things came about: humans were created by God and soon (through the disobedience of Adam and Eve, in the main interpretation) became mortal and immanent, like everything in nature. But God's version of humans remained fundamentally different from the rest of Creation in that they possessed one faculty which linked them to the transcendent world: they have another 'reality' in them, they have a will (linked to the soul, mind, conscience or however it was specified in later historical analyses). In many interpretations this 'second reality' is eternal and may live forever in a transcendental world, after the perishable natural part of a human being would have died (certainly in Christianity and Islam). Along this perspective the focus lies on the essential difference between humans and the 'rest of nature': humans in that ontology are in an absolute or essential sense different from the rest, which can be called nature, because of this 'godlike' aspect. They are different, not just another shade of diversity. Hence, humans are allowed (by God in Genesis) to use or otherwise reign over the rest, which lack that distinguishing essential element of a will since they lack a soul. Generally speaking, there emerged an agelong agreement throughout the area where the 'religions of the book' gained fundamental influence, the deep intuition that humans had such distinct faculty (consciousness, soul, etc.) which implies the capacity to choose and hence to have morality and/or religion (Descola 2016). All of this, moreover, would be transferred through learning procedures. And all of this was – and, to a very large extent, continues to be – believed to be absent in the rest of nature, or maybe to exist in a very reduced way in animals. Church rules such as the training of human beings in this ontology (through religious education), of regulating behaviour through rules for humans as in principle distinct from those for the rest of nature, even the obligation of believers (especially Christian and Islamic believers) to convert all humans and never convert any being of 'the rest' – all this testifies to the power of this ontology through the ages. Even today, it is considered soft if not outright ludicrous for a philosopher to claim that the hierarchy between humanity and 'the rest' is ill-founded and that animals reason, or that plants organize themselves in a community of communication and interaction and can even become 'colonizers' over other plants to some extent. (Very recently, some researchers started along this road – but they are rare, and it took five centuries to dare and explore that path.)

A by-product of this ontology was (and is) that for centuries non-western so-called 'primitive' people – not merely their thoughts or products – who were 'discovered' by western seafarers could not be fully adopted in the section of 'humans' of that western ontology. They were termed 'wild', or 'part of nature (instead of human in the proper sense)' and theologians took centuries to discuss whether or not these 'others' belonged to one or the other part of Creation, and hence should or should not be baptized. Even when these others were granted a place within humanity at some point, they were (and still are) treated as different in the sense of not civilized (right up to the Second World War), or 'less-' or 'un-developed' since the start of the UN daughter organization of UNDP, focusing explicitly on their development to become more 'like us, westerners' (Fitzgerald 2021). Present-day racism and cultural-identity thinking stem from that 'evident', 'undeniable' dualism, to my mind.

At the time of the first humanists this blind adoption of European (and later western) superiority was unquestionably present in their views as well. Only very rare exceptions can be found, for example with the remarkably open mind of Michel de Montaigne. Living at a time of fierce religious conflicts between Catholics and Protestants lasting at least two centuries, he expressed his doubts on European superiority when coming into contact with indigenous people who were put on show at the courts of his time. He wondered how Europeans could kill each other mercilessly in lengthy, atrocious religious wars and still claim that these 'others' were part of nature, i.e. they were wild. But Montaigne is a lonely source of such open-minded thinking (see, e.g., chapter XXXI in Montaigne 2004, originally 1595). This is, apart from everything else, one of the reasons Claude Lévi-Strauss (2016, originally 1992) pointed to Montaigne as the most important French philosopher, to his mind.

Happily enough, in our days David Graeber and David Wengrow (2021) went through the sources of forgotten voices of 'others' in the first centuries after the discovery of the Americas. But the authors had to stir up and remind the contemporary citizen that the voice of these 'others' has been silenced for centuries, that is since they were characterized as savages, primitives and so on by the political and the intellectual elites of later generations.

Since Graeber and Wengrow (2021) is becoming a bestseller, it is good to remind the reader that a few recent anthropological works have been questioning this dualism in western ontology for a generation now, with the result that the dualism begins to be recognized and problematized in a more systematic and straightforward way. A shift is being triggered by developments in the political discussions now, which are fed by Native South Americans. In recent years, the eminent French anthropologist Philippe Descola (2005) made a synthesis of European–western premises in the dualistic ontology I discuss here. The general reasoning runs as follows:

- only human beings are seen as possessing a mind, with its 'internal' sensitivities like self-consciousness;
  hence, only humans can develop morality;
- from this, it follows that only human beings can be responsible for their deeds and their thoughts;
- from which follows that only humans can have rights.

It is obvious that humanism (and, later, modernity) did not question or in any other sense go beyond the premises of this view. This ontology was accepted by people without criticism as correct and sufficient. In subsequent centuries the emphasis shifted partly to knowledge, and also the modernist of today will mostly take for granted that, with the enormous growth in efficiency and impact of contemporary sciences and technology, we can automatically justify the western shared ontology underlying them. From this attitude it follows that the insights, knowledge and even ontological ideas of other, so-called un-developed cultural traditions need not really be taken seriously. Or, if they are showing survival value, they will probably in a very basic way sit on the same line of reasoning about 'nature' as the westerner, it is believed. This may strike the reader as a surprisingly general statement, but by way of example any student of the anthropology of religion or comparative religious studies will tell you that this is the fundamental 'stumbling block' in such disciplines: such deep a prioris of an ontological nature cause insurmountable barriers to approaching the recent interdependence mentioned earlier.

Looking closely at the worldwide situation we have landed in (with the crises pointed out in chapter 2) it is high time to slow down and self-critically look at the ontology we have been taking for granted all this time. Indeed, in a general sense, one can only see a continuity of ontological premises in our, European–western history. Since the Mediterranean heritage European–western people hold the conviction that humans share some natural elements with other species (mostly the higher mammals), but nevertheless exist as if on a fundamentally distinct layer of reality: Adam was told that humans are different from the rest and should somehow reign over the rest (nature) and take from them what he needed. Throughout the time of Christian (and Islamic) supremacy these premises went unquestioned, and they were adopted basically unamended by humanists and modernists; some values and social constructs of the theologically dominated worldviews were occasionally heavily criticized by them, but not the ontology. In the past three centuries capitalism (with its unquestioned value of private property: Piketty 2019) used the more efficient tools of science and technology to expand this value orientation on a worldwide scale. In no time everything and everybody was repositioned in terms of the supremacy view of the westerner. Throughout this period humanists sided most of the time with a superficially de-mythologized view of reality, and thus did not question its ontology in any deep sense (I leave aside attempts at criticism such as Heidegger's, since in my understanding they always remained within the European perspective).

Recently some worries surfaced about the way we treat others, as well as nature – for example, philosophers would advocate that animals should be granted rights (Peter Singer and others). Notwithstanding these attempts at recognizing the rights of non-human parts of reality, the more fundamental attitude stayed intact: if I were to propose granting rights to water; air; or, more concretely, even the North and South Poles or the deep sea, I would risk becoming a laughing stock for many and would be classified as the enemy of those powerful economic groups which are 'harvesting' the oceans or 'making accessible the underground' of the polar areas. I would certainly land as 'an enemy of progress' in the eyes of many. In a general way, therefore, the dualism between humanity and 'nature' in the ontological view stays largely intact.

Still, some anthropologists started insisting on deeply alternative ways of thinking about humanity and its possible relationships with all other phenomena, detailing such claims with painstaking long-term research in different communities. Looking at fundamentally different, so-called 'horizontal' positioning of other aspects of reality (or nature?) vis-à-vis humans, several cultures or peoples we have classified as 'primitive' or 'undeveloped' in past centuries are shown to have respectable and possibly life- or earth-saving intuitions, which are profoundly different from the dualism taken for granted by the European–western conqueror.

These anthropological findings are particular and were often filed away as 'exotic' or even as 'curiosities' about the human developments of 'past' communities. In order to grow out of the supremacy attitude, anthropologists started suggesting that we need to develop a way to look at humanity from a genuinely comparative perspective. That is to say, we need to allow for different perspectives in knowledge and even in intuition (the ontological level). This is less easy than it sounds: how can we take diverging and apparently conflicting insights on humanity, on nature and on the many possible relationships between them seriously? Does this not force us to reconsider age-old views on truth or even on reason – and invite us, for instance, to interpret in a more modest way the self-satisfying rule of logical consistency, as an outgrowth of the superior status of humans as isolated from the rest of nature? Indeed, can the dominance of context-independent logical reasoning be taken for granted when humans and the rest of nature are thought of as horizontally interconnected, rather than hierarchically disconnected? And continuing along this line, should we then rehabilitate the ancient sophists instead of disqualifying them as relativist super-pragmatists as generations of Platonists, theologians and positivists have done (Perelman and Olbrechts-Tyteca 1957)? Should we not reconsider critically and earnestly the ontological premises we have built our knowledge on? We could start with taking seriously those 'eccentric' thinkers in our own history, such as A.N. Whitehead and his 'event'-philosophy (Whitehead 1919) or some Buddhist-inspired scientists of more recent date (D. Bohm for one)? This is not an appeal to relativism, but rather an invitation to critically examine the pre-knowledge premises we have taken for granted since the European scientific revolution started. With a broader and possibly a pluriversal perspective on such basic intuitions as what is human, what is time, or why 'to be' would be either a more or a less valid premise than 'to happen' or 'to connect'? From there on, we might prove that other cultural traditions have developed alternative intuitions ('ontologies') and their intrinsic value, which we westerners have been cancelling out over the past centuries without even looking at the possibilities they might hold for another, more durable or more encompassing perspective for survival.

When we open our mind to such an endeavour, we run into a small, but growing group of scholars who try to develop a comparative view on human traditions of thinking and doing in their proposals for survival. In anthropology, this is now dubbed 'the ontological turn'. Recently, Descola (2005 and 2021) worked along these lines and he is gaining influence in the discipline of anthropology and beyond. Tim Ingold (2004, 2017), myself (Pinxten, van Dooren and Harvey 1983 and Pinxten 2001) and some others started, along independent tracks, work that shows strong parallels with Descola's programme. But most of all, an ever-growing group of South American scholars has been developing a postcolonial critique which digs deeper than the 'internal critiques' of westerners and often links up with the ontologies of the indigenous groups from that part of the world.

To give one concrete example: the ontology of the Quechua will hold a category of 'Earth Beings' which includes mountains, forests, animals and much more next to and interacting with humans within the one 'horizontal' world (de la Cadena 2011). Such phenomena or 'beings' are treated respectfully by the Quechua, since they 'people' the earth alongside and in close interconnection with humans. Ontologically, each of these groups of 'beings' is distinct from any other one, but they are all deeply interconnected in the holistic ontology of that tradition. I encountered similar ideas with the Navajo Native Americans: notwithstanding almost two centuries of White American dominance (with schooling, law enforcement and suchlike) people kept telling me that speaking and acting (ordinary acting or ritual performances), but also thinking should be understood as ways of dealing with other phenomena. In this ontology, human thinking–talking–acting always involves the impact of all phenomena on everything, and hence on people. So the person who positions himself as detached from and maybe above the rest of reality might almost certainly cause harm to other phenomena and/or to himself without even noticing it at first. In that tradition shamans could determine harmful behaviour and its impact on everything, whereupon medicine people would try to correct or 'cure' the disequilibrium by holding elaborate ceremonies over the 'patient' (Witherspoon 1977; Farella 1984). Within the holistic view on reality of this North American tradition, a similar yet differently articulated perspective on the interconnectedness of everything is to be found.

In Descola's synthetic overview of similar ontologies in diverging parts of the world, it becomes clear that we have a treasury of knowledge, but also ritual and daily practices in many non-western cultures, which allows to recognize or build quite different relationships between humans and other phenomena: exploitation, let alone the pillage mentality of late capitalism is absent from these so-called primitive traditions. In an era when the crises westerners have brought forth by their approach to the rest of the world, sited in and justified by their dualistic ontology, the least humanists should do is self-critically look into this ontology and maybe let themselves be inspired by these non-western traditions for alternative views. As I interpret Descola's work, it aims at identifying at the very least four different 'ontologies' or deep intuitive approaches in cultural traditions around the world: the animistic, the totemistic, the analogistic and the naturalistic view. In the most recent work of the author, this quartet is also used to understand the different ways artistic (and religious) 'figuration' (representation) has been growing. The ways humans see themselves in connection with other phenomena, as well as the types of impact phenomena might have on each other (humans but also other 'beings') and the forms of agency found should be recognized and studied properly (Descola 2021). Also, the different intuitions on time, on existence and on form and figuration developed by different cultures around the world should have the attention of anthropologists and philosophers.

When we integrate all of this material into our general scope of humans-as-part-of-reality, it is clear that the ensuing picture of reality will be much richer and might have more potential for survival than the sole, atomistic, context- and time-independent worldview we have been working with in the West over the past few centuries. This does not deny that the angle we have been developing has been successful in the treatment of particular questions: the study of matter, and to some extent of living creatures, has been substantially expanded in recent western history. Some branches of natural sciences and of medicine have progressed remarkably. But, as Nobel laureate Ilya Prigogine kept repeating (see, e.g., Prigogine 1969; Prigogine and Stengers 1984), the time-independent intuition on physical matter has reached its limits in the sciences, and thus basic intuitive or ontological premises might come up for critical examination.

In similar critiques (in domains such as biology; medicine; and, most of all, in some social sciences), the narrowness and bias of our western exclusivist perspective on phenomena progressively hinders an open-minded search for knowledge which would be sustainable and decent at the same time. Gender discussions, decolonization critiques and anti-racism actions abound. Finally, the insistent and growing crises we are confronting in the present era (such as climate disequilibrium, the destruction of biodiversity and growing inequality, to mention just three of them) urge us, in my opinion, to earnestly work on the shift towards an open-minded critique of the limitations of our worldview, however successful it has been in some domains of human life. It is in that respect that I understand the so-called 'ontological turn' which, among others, Descola's work is promoting in the West. Sure enough, major discussions will have to continue and to be deepened: to what extent can such a proposal be sufficiently free from western biases (see the Ingold–Descola discussion of 2016)? How can we reach enough sophistication and, hence, evidence-based knowledge when we expand the intuitive premises? And so on. However reticent we may be, the chances are that we will have to engage in painstaking work in an honest way and with more modesty and caution than in the past centuries, in order to open up our perspective to guarantee less destructive and more durable responses to the survival challenges we have to meet. That is, put in an extremely generalized rendering, the invitation I extend to humanists. It sketches the scenery for what I call here NEED humanism.

More Examples of What Seems to Be Changing in the Wonderful World of Science

Before I explain the 'programme' behind the NEED-word, it is important to go into at least one example of what is meant by the 'ontological turn' in anthropology, and to indicate that in other sciences similar 'soul searching' seems to be starting.

I referred to Descola's earlier discussion of the 'ontological turn' in some Latin American anthropological research. The inspiration in that area of the world came from Indigenous cultures and their emphasis on the horizontal interconnectedness of everything in reality: in a fundamentally holistic view of things, they see the differences of all phenomena vis-à-vis each other but at the same time the deep interconnectedness. What humans think, say or do impacts on animals, trees, water, sky and whatnot, and the other way around. Whether or not this should be seen as a form of 'ecological thinking' is a subject for academic debate (see Hitchcock and Galvin 2023). What matters is that the hierarchy between the human 'God's Eye View' (Pinxten 2010) on everything else, and hence the mandate to use 'nature' at will, is absent: humans in those traditions do not see themselves as detached from the rest of reality since they do not 'look over the shoulder of the God-Creator' at reality. In the Native American view humans remain immersed in the totality of reality, in an imminent view on everything. From this it follows that they adopt an attitude of caution or reticence: since humans are part of everything, they will most likely not interact with other parts of the totality without expecting some impact on themselves. In the course of my own fieldwork Navajo people kept telling me that they never 'erase' any animal or plant of a species in their environment, the way 'white people' taught them to do – cleaning the land of all the so-called wild plants and hunting down and eliminating all animals that might feed on your crops. Navajo kept saying that this is not wise, and lacked understanding of the context humans live in and live from.

The example from Quechua culture as described in the work of the Latin American anthropologist Marisol de la Cadena (2011) on 'Earth Beings' makes this more concrete, I guess: the 'ontological primitive' (in philosophical jargon) is the Earth Being, and it is manifested in particular mountains, forests, humans, animals, plants and so on. They are all instances of 'Earth Beings'. Together, in the complex structures of interconnectness in which they exist, they constitute the one reality. Obviously, when multinational corporations discovered profitable natural resources on Quechua territory they started to approach their 'leader' in order to have him sign a contract for exploitation. The religious–social 'leader' could not go along with the industrial partner, since this was both 'unnatural' and an act of treason against his own people and their worldview or ontology. He went underground and thus stopped any progress by the multinational for a while. When the latter insisted the national government yielded and the Quechua territory was taken through a military plot: people were chased away and land was destroyed in the name of economic progress for the region. The battle is not settled, though, but so far the Quechua local groups are victims of the pillage mentality of the multinational group, which justifies itself by the presumed superiority of its economic theory on progress. The arguments of the Quechua draw to a considerable extent on their holistic ontology, which the 'leader' defended as long as possible; when his strategy in negotiations failed he had to flee. The dualistic view of the multinational allowed for this land grabbing, while the holistic local one was for the moment considered to be a remnant of 'primitive' knowledge. Similar examples have been documented in Brazil under the government of the authoritarian former president Jair Bolsonaro (with disastrous deforestation and attacks on biodiversity, and murder gangs dispatched against Indigenous peoples), but also with examples from the newly accessible polar area in Greenland (Ingold 2017), the formerly colonized Congolese territories, climate disaster in contrast to the Māori holism view in New Zealand and so on. Whatever else, the western heritage of humanism is only selectively, if at all, referred to in this reckless pillage by western globalized capitalism and (in a somewhat different way) of present Chinese imperialism with the 'new Silk Road' (Frankopan 2018). This more political and activist consciousness, fed by recent anthropological findings, provides another argument for me to launch the said reset of humanism as a NEED humanism. But there is more.

Other Views on 'Objectification' Are Emerging in Scientific Disciplines Today

Anthropology has always been a discipline almost in the margin of so-called 'serious' sciences: we were focusing on non-western cultures exclusively in the first decades of the discipline (1850–1920), and we chose to develop qualitative approaches and even questioned overly easy quantification. Anthropologists also went in for multi- and interdisciplinary research, which was considered to be weak and hence rather irrelevant till only a couple of decades ago. With the globalization of capitalism and the subsequent interdependence of humanity the status of anthropology starts shifting: such deep, qualitative issues as meaning, values, even different worldviews (e.g. holism versus dualism) begin to be recognized as relevant. The novels I referred to in the preface to this book address these issues today for a much larger audience: it seems impossible to me that a book on trees and on how a growing group of idealists quit the rat race in order to live and fight for a holistic worldview – where forests are seen as communicating communities rather than just objects at the mercy of the supremacy of humans – would be published a generation ago. Now, it allows its author to contribute to a public debate on sensible and sustainable alternatives to the pillage mentality humanity has been exhibiting (Powers 2018). Richard Powers' book is a bestseller, which at least shows that the horizon of political consciousness and ethical choices is changing.

However, next to anthropology, in several sciences the 'obvious' dualism of humanity–nature seems to be progressively challenged. In the traditional, dualistic way any part of 'nature' (in and outside of human beings themselves) can be objectified. The most popular approach, then, is that of reductionism: anything can and should be split into its constitutive parts. These parts are simpler and should hence be easier to study – e.g. the biological processes of cells should be understood best by looking for their chemical (and ultimately also physical) components. The sum of knowledge about the parts then allows for a relevant, or presumably full or sufficient, understanding of the larger entity (i.e. the cell). Without doubt this approach is very powerful when studying relatively simple, and most of all 'dead', matter. When biology became more and more the 'queen science' (since the 1960s with the emergence of DNA research, for example) reductionism started being challenged systematically: the greater the number of complex 'entities' that are the subject of research, the more problematic a simple reductionist approach becomes. Prigogine's claim, which I mentioned earlier, announced this shift in thinking: the end of atomism became a possible and maybe deeply relevant road for research (Prigogine 1969). Ecological thinking, generalized into broad systems thinking, tipped the scales in a more principled way. The laws and features of the more complex level cannot be adequately understood by simply referring to the sum of knowledge gained on the constitutive parts.

In the same trend some social sciences (especially anthropology, I should say) started studying persons in their contexts: the social, biological, political and historical context. Focusing on such broad and complex subjects will increasingly preclude easy, and easily quantifiable, analysis – for example, the quantitative 'measuring' of cognitive capabilities through quantifiable test results in the typical Programme for International Student Assessments (PISA), whereby children worldwide are tested on their mathematical capacities. The tests are recognizable for any psychologist trained in quantitative research. When scrutinizing the questions, though, it appears that market-conformist thinking is dominant in most of them and no other worldview is present in the test battery. In a rather straightforward way, mastering market thinking acts as a hidden selection criterion in the OECD's tests (Standaert 2021). Thus, when testing children's responses the psychologist implicitly (and maybe even unconsciously) presupposes that an exclusive set of features of market thinking can be taken to be universal: choosing what would be 'the best deal', and going for more instead of what one really needs, is presented as the 'clever' choice; calculating in terms of money and monetary gains or losses is more 'advanced' than other choices in these tests. By showing a lack of interest in, and maybe even consciousness of, contextual parameters other than 'quantifications', this regularly organized research by the OECD misrepresents children who are raised in other contexts. Not surprisingly, they are diagnosed as lacking in cognitive competences. Moreover the OECD, the patron organization of PISA research, prompts governments around the world to adapt their schooling programmes according to the results of PISA research, on the basis of which the countries are ranked as more or less economically progressive or backward. Thus, this type of reductionistic quantifying research in fact becomes a political tool, pushing people to become more exclusive market players 'in the name of science' (Pinxten 2016, Standaert 2021). Even while allowing that the researchers involved are not evil-minded, one can only remark that the consequences of this 'objectifying by reduction' (i.e. simplifying by leaving out diversity) is highly questionable. For one thing, it is at best the use of one element of humanism and Enlightenment (free research, science instead of common sense) in a decontextualized perspective in order to push for very debatable ends, looked at from a broader humanistic perspective.

In evolutionary theory the debate has picked up these kinds of critical notes when shifting away from simplistic nineteenth-century mechanistic thinking towards systems thinking – e.g. Donald Campbell (1989) sketched a new pyramid of complexity of learning forms. At each subsequent layer ever-more complex and less deterministic types of adaptation, and thus 'evolution', obtain. In his 'hierarchy of learning strategies' Campbell looks at elementary forms of living beings at the ground level (the amoeba) and 'moves up' gradually to understand how ever-more complex and less deterministic forms of learning can be discerned – from animal instincts; over stimulus-response forms; learning-in-social-context forms, such as in Lev Vigotky's theory; to self-critical scientific and philosophical learning; to artistic learning and 'culture' in the broadest sense. Each 'higher' level – that is, each level detailing more complex phenomena than the lower ones – has features that cannot be reduced to those of the levels beneath it. In that view, in order to understand properly and develop relevant knowledge about complex phenomena the researcher will come up with new or seemingly irreducible features and processes in ever-more complex contexts in which these phenomena exist.

Campbell was a psychologist with philosophical inclinations. But similar ideas and theoretical proposals can be found in natural-scientific approaches on evolution: I think of chaos and complexity theory (see, e.g., Kauffman 2010) and the very productive line of Latin American evolutionists since the later decades of the past century (Maturana and Varela 1980, and onwards). Not surprisingly, in line with these studies botanists and other natural scientists are exploring what a 'sociology of forests' would be like, and how dolphins and whales – as well as bonobos, wolves, elephants and other species – are part and parcel of horizontal 'webs of existence' (or the 'chain of being', or Gaia or whatever other holistic concepts are discussed now) together with humans, and belong in an interconnected world of 'nature'. To be sure, there is much more critical discussion going on today, but I mention these examples only to indicate that a deep reshuffling might be under way: general systems theory, founded by older thinkers like Ludwig von Bertalanffy, seems to be alive and kicking, I suggest.

Together with the growth of systems thinking I perceive an increasing influence of theories on self-organization. Where original evolutionary theory had an almost exclusive interest in the 'biological' processes at species level, a more recent branch allows more room in the theory for self-organization within species living in environments. Concepts such as epigenetic evolution, complexity and self-organization were coined in order to make the original theory less mechanistic and deterministic on the one hand, and to expand the exclusively genetic view with such complex phenomena as community or neighbourhood in the development of plants and animals, and of humans, on the other. That is to say, it is not odd – or, at the very least, it is acceptable – today to think of the sociology of trees in a forest, linked by networks of fungi and moulds which have the trees communicate and interact with each other and with the other species in the environment. Also, definite policies for national parks are shifting towards respecting this 'community' view on ecology and evolution – e.g. the programme of Yellowstone National Park, where the wolf has been reintroduced and shown to be able to restore the total ecological equilibrium which had been seriously disrupted by its earlier extinction. This does not run counter to genetic theory and evolution. Rather, it broadens the latter by allowing serious consideration for epigenetic factors. A forceful and more elaborate theory in that line of thinking came out of the so-called Santa Fe Group, notably, Stuart Kauffman (2010) and his collaborators worked on a theory which could explain the complex systems one finds on many levels of organizations. Not only could he (as an organic chemist) determine complex processes at the level of cells, but he also worked out an intriguing theory whereby processes of self-organization are added on to more basic levels of organization in order to explain such complex and evasive levels of existence as culture, religion and art in a coherent frame of reference that reminds one of Campbell's earlier attempt at a hierarchy of learning strategies. As a natural scientist, Kauffman makes great efforts to avoid simplistic, let alone reductionist, statements and thus invites scholars from different disciplines to collaborate on a broad systems approach for the most complex phenomena we know of.

In my view this allows for a major step away from an exclusivist, and hence rather parochially biased, attitude which has been planted and nursed in the mind of the European–westerner since the 'discovery' of the Americas and beyond (about five centuries ago). It probably took a broadminded researcher such as Graeber to pursue the critical analysis in a free and courageous way. In the aforementioned book (Graeber and Wengrow 2021) it is shown how the critical voices the colonial settlers encountered with the Native Americans were mostly repressed over time. Recently, such voices have been sparingly rediscovered: to mention just one example, Laura Nader brought out a volume (2015) with a series of reflections by mostly Asian and Near Eastern 'others' on European–western ways, some of them dating from centuries back and neglected over the same time.

In a former book Graeber (2013) had already emphasized how some Founding Fathers of the American Revolution (namely, Thomas Jefferson and Benjamin Franklin) showed deep knowledge of and honest respect for the 'federal model' of the Sioux Nation, from which they adopted not only the icon (the bundle of six arrows) but also some organizational principles. This fact is proven from the diary material Graeber investigated. However, these 'Indian roots' were immediately covered up by the story that the federal model of the USA would have been inspired by the Dutch system. Graeber shows clearly that this is false, but the need to justify oneself

54 • Humanism Revisited

through reference to the 'civilized' world was so strong for the American Founding Fathers that the Native American origin was suppressed. Also, the Sioux were not exceptional in this regard: the Iroquois, for instance, developed a 'league' which in several respects can be considered to be a federation as well. A 'founding father' of American anthropology, Lewis Henry Morgan, brought out a well-documented study of this political formation almost two centuries ago (Morgan 1851). His work never reached a wider public, though.

Both the shifts in evolutionary thinking and the rehabilitation of 'the other' can inspire us in today's predicament, I suggest.

Summing up/a Comparative Project

If we follow the path sketched above, we can now turn back to Descola's grand scheme and look from a comparative perspective at what humanity has elaborated at the level of ontologies, or 'intuitive understandings' ('worldings' – *mondiations* in French: Descola 2021). He claims that the Enlightenment idea that one universal nature exists, of which different cultural traditions give ever so many partial versions, is biased, since it recognizes 'other' worldviews at the most as a thing of humanity's past. They are understood as a variation within the one dominant and uniquely 'true' frame of reference, which is the western view: the latter is the most sophisticated or adult version. In the present era, however, we run into tremendous problems with this view on things, and we should at the very least become more cautious and less self-confident. This critical remark focuses on the intuitive foundational insights rather than on the procedures of investigation 'applied' from them. Of course, the scientific attitude in principle built in checks and balances and even notions of caution. Even so, the recent organizational turn towards the 'corporatization of academia' (Sahlins 2008) denies some of these important principles: patents, profits, competition are becoming primordial values, and colleagues from the natural sciences assure me that this tendency is deeply changing their practices in research – e.g. seeking funding for control or repetition research is more and more difficult today, since checking the validity of research is often too costly or even deemed unnecessary (leading to scandals of 'fake' articles published even in high-quality journals, but also the marketing of new science journals that will print almost anything against cost, etc.). With these trends and the recent explicit attacks on scientific fact production by important political, media and economic players (the trend of 'alternative truths') added on to the crises I have been talking about in this book, the time seems ripe to look again at the presumptions we have been holding.

Independent of these critiques I cite Descola's suggestion to look at the underlying tenets of research programmes critically once more: instead of stating that a series of approaches, which partially cover the one 'reality' we project as objective and somehow detached for the western (objectifying) observer and as superior to all other perspectives on reality, we should look more cautiously at the apparent diversity of cultural traditions (cognitive, religious or interactive ways of dealing) as 'une diversité de processus de composition des mondes' ('a diversity of processes of composing worlds': Descola 2021:10). From this shift in attitude, it follows that all 'ways of worldmaking' (to borrow Nelson Goodman's terminology: Goodman 1978) become feasible. They will start from diverging intuitions or assumptions which may differ between them, share elements or insights, use 'stepping stones' or insights from neighbouring ways and so on. From this perspective, a unique, correct, superior or altogether 'right' way that will disqualify all others as intrinsically false, irrelevant or 'primitive' is unlikely. Instead, it is important to consider the situatedness of any 'way' vis-à-vis a particular context (desert versus fertile soil with enough moisture, for example, but also historical human parameters such as type of barter or trade, religious or political regimes and so on). It is important to understand that this is not a relativist position, claiming that all and anything is as 'true' as anything else (Restivo 2021). Indeed, we are not dealing with the epistemological level here (where conditions of truth and falsehood apply), but rather with the ontological one: we are focusing on 'worlding' or, in common terms, the intuitive level out of which so-called cognitive or properly epistemic assertions or models and theories will be conceived to be tested and applied or not. In the course of his thorough and very elaborate analysis Descola (2021) shows, for example, how the analogical and hence, in the naturalistic and later scientific 'worlding', rejected view on perspective-directed-towards-the-perceiver is dominant in the Christian worldview. This perspective entails that in the drawing or the picture everything appears as if seen from the point of view of God. Under the influence of several contextual changes, such as the growing role of manufacturing and trading individuals in the emerging cities of Italy and Flanders during the Renaissance, this way of viewing was left behind or turned around to become the linear perspective. Here the perceiver is the mental organizer of the world. Or, put differently, the world is depicted in art and in knowledge treatises henceforth as if looked at primarily through 'the eyes of the individual human being'. This way of looking and 'worlding' is a clear prerequisite and hence a stepping stone to come to a naturalistic way of worldmaking, the intuitive basis of the scientific worldview to come. However, the analogical 'worlding' remains present in certain domains (e.g. religion).

The reader will now understand how I try to look at humanism (and the Enlightenment): it should be understood in relationship to the context in which it was lived. This context holds a way of 'worlding', but also the physical and political–economic conditions of a community. Again, this does not disqualify humanism today as 'something of the past'. However, it does make clear that humanistic values and presuppositions might need a critical reset if the contextual conditions change considerably. In my view, this is indeed what has happened: very briefly, the world is interdependent today and the hegemonic power of Europe in the preceding centuries is largely gone now, although the impact of conquests and the economic globalization of the past will continue to co-determine the fate of the world to a large extent. What kind of humanism emerges, as a way of 'worlding' with a particular notion of mankind, of exclusion/inclusion and so on? How does it differ from the original?

My claim is that the way of reasoning of the humanist may have powerful potential to cope with the present world predicament, provided we update it so as to become a way of 'worlding' which speaks to the present situation of humanity and the world. For one thing, the clearly fashionable focus on the individual at the time of its conception is perfectly understandable in the context of the time of Erasmus and Montaigne, but it has, I claim, run itself into the ground today in its early and simplistic form – i.e. that independently thinking and acting individuals can be seen in themselves as the sole 'measure of everything', as the old humanists claimed. However, the individualism which was founded on this premise calls for serious disruptions on a global scale today. Individual freedom of thought might be a great value, but what about private ownership or other sovereignty values today? In an interdependent world, what if these are uncritically held onto as superior to anything other traditions have thought up? My proposal is that rethinking these values would focus first and foremost on the different ways humans are known to relate to reality. Concretely, such an endeavour could start with evaluating interhuman relationships – i.e. as westerners go beyond the Eurocentrism of the old view. This is precisely what a comparative perspective on the humanities and the social sciences is aiming at.

But there is more: that is what the acronym NEED covers. The full version reads: NE for Non-Eurocentric, E for Ecological (stretching from biological to climate processes) and D for a Durable or resilient approach in a general sense (implying also political–economic long-term reasoning).

Chapter 4

NE

Non-Eurocentric Humanism

Presuppositions

As a slogan, I proposed: humans have a high degree of 'dolphinhood'. Or, to put it differently, humans are 'super-collaborators' (after the title of Van Duppen and Hoebeke 2016): they are social animals, with empathy and even a sense of helping others without expecting direct profit from it, beyond identifiable self-interest. In order to appreciate this, it is good to look at the three types of actions humans are capable of. We can act vis-à-vis our own person, we can act in person-to-person or face-to-face relationships in groups and we can act in virtual groups like communities (without person-to-person contacts). These are three different types of action, defining the social dimension of humans.

Human beings are physically and mentally separate, individual beings: they each have their own body and a consciousness of this and (to a degree) of the non-physical aspects of themselves, with the choices, feelings, preferences and so on that come with all this. Human beings can obviously perform actions that are aimed at this individual unit: eat, clothe, learn, muse, right up to harming or killing this unit.

Humans also have face-to-face contacts with each other. I act and react to other persons: I speak to them, look at them, touch them, and develop over the years a series of types of interactions going from superficial and public to quite intimate contact with members of my species.

In the third place, and at yet another scale, I develop still another type of actions, which take the form of virtual interaction with others: I can feel at one with lots of people I have not and will never meet face-to-face, or I can hate groups that I have never run into physically, neither heard or saw, in my life. I take actions that have me known as a man, a westerner, an intellectual and lots of other things and become part of minorities or majorities that influence the organization of society, future generations maybe, and so on. This, again, is quite a different sort of action and interaction than the two former ones. The impact of this type and level of action may become more striking in the present era, where purely virtual 'communities' interact to great effect in the political evolution of states – Facebook groups and suchlike, but also Amazon or Alibaba networks, are beginning to shape public life to a great extent now.

The types and impact forms at any of these three levels are expanding or shrinking today. It is relevant for humanists to at least look at the differences between today's habits and formats in these matters and those of the Renaissance; humanists back then emphasized that the role of the individual should be expanded at the expense of that of the other levels (the face-to-face groups of nobility and clergy, for instance), which had to be countered and limited. A point of attention for the humanist in our era should be the issue of the ideal role of the three levels today. Simply more power for the individual level, and more freedom of each individual to do what she wants to do might cause fiercely anti-humanistic effects, for example. Think of individual rights to oppose vaccination in the interdependent world of today, or the sanctity of the individual right to cheap flights and cheap consumer material (produced by children in cheap-labour countries) etc.

Alongside social contexts, human beings are raised as knowledgeable, ethical, aesthetic and spiritual beings in cultural contexts, where all that is transferable through learning processes turns the human animal into what we call a 'nurtured' individual. Through these learning styles and content, we adopt tastes, values, beliefs and other conventions – and we learn certain rhythms, notions of time and of love and hate.

In his magnificent overview of the neuroscientific–psychological–socio-cultural makeup of humans, the neuropsychologist Robert Sapolsky (2017) sketches a remarkable synthesis of our present knowledge of these processes: there is a genetic basis for much of what makes human beings human, but there is a large margin (and less determinacy) where the human being is shown to be pliable, creative or self-organizing to an extent. We share the existence of such a margin with many other species, but Sapolsky closes this grand book with the remark: 'We're the only species that institutionalizes reconciliation, and that grapples with "truth", "forgiveness", "reparations", "amnesty" and "forgetting"' (Ibid.: 638). In my view this remarkable finding on the institutionalization of such features of the third level of action holds an invitation for today's humanists to reflect critically on some old principles: we live in a time where on the one hand e-control systems are being installed worldwide at a tremendous speed (from face and finger-print recognition to e-banking, etc.) and on the other hand 'greed is good' and populism (in the name of individual freedom: see the storming of Washington's Capitol Hill on 6 January 2021) become rampant.

The Museum of the American Indian

I start from a concrete example. Visiting the marvellous Museum of the American Indian in Washington D.C. (www.americanindian.si.edu) makes clear what I am heading at. This unique and new museum was thought out and constructed according to the explicit Indigenous indications on forms and functions advanced by several groups of Native Americans. For one thing, there are no straight lines or straight angles in the complex and a small river runs through the ground floor. The museum confronts the visitor with a permanent collection where appalling images and statements of a not fully ended colonial history are shown. I heard several US visitors making remarks of disbelief during my visit: did our government and did the missions really act so atrociously? More than one visitor was said to be ashamed of the treatment that was reserved for these 'others'. The exhibits, sitting on the Mall and thus in the power centre of one of the world's superpowers, show how the exploitation and even annihilation of a good part of the original local population led to present-day poverty and discrimination for most of the survivors of these cultures. The missions, claiming to work for the betterment of these peoples, played an especially devastating role since they aimed at changing the Indigenous humans in their very cultural core: their ethics, their beliefs, their notions of humanity. The shock the museum exhibition produces, I guess, can best be appreciated against the background of decades of brainwashing by the media, where the 'hero-cowboy' was always depicted as punishing or reprogramming, if not killing, the stupid, filthy and unreliable Indian. The former was the hero, who represented a superior culture and who was aiming at ameliorating the 'wild country' in the name of progress and civilization, whereas the Indian was obviously stuck in a former, primitive stage of humanity. Overall, the missions did not run counter to this appreciation of the superior westerner but offered a more systematic approach to deep reprogramming, even if this led mostly to alienation with the Native American people. In my fieldwork from the 1970s onwards, I cannot say that I encountered any other attitude among missionaries on reservations: their reaction varied from mild tolerance of unorthodox beliefs (e.g. of the popular peyote cult) in the Catholic mission of Lukachukai to a repeated refusal to help when I shared my troubles with Navajo neighbours (getting stuck in mud, and even having cut my feet with an axe, I only met with refusal by a local Protestant mission 'because none of you belong to this church'). All in all, the general line seemed to be that the 'Indians' had to be saved through religious education and to shed their native views along the way. The attitude of superiority in the 'White' missionaries which Graeber and Wengrow (2021) criticize was omnipresent: 'they' belonged to a former stage of human development and 'we' were at the very least one step ahead of them on the unique track of progress for humanity. In Graeber and Wengrow's analysis this conviction became dominant rather late (with Jean-Jacques Rousseau's 'noble savage' and, from an economic perspective, with Adam Smith's view on primitives who lacked the progressive notion of private property). It was launched as a reaction to original Native criticism of European pettiness and harshness by Native Americans. However, those critiques were lost to sight – after an early, short period of popularity – when the imperialistic, capitalist policy in the colonies began to determine political horizons (generally speaking, certainly since the American and the French Revolutions). In the early twentieth century the new folktale of 'cowboys and Indians' shaped the worldview of generations through books (e.g. those of German author Karl May) and of course through the film industry. The museum in Washington, DC addresses these issues head-on and shows how the survival conditions of Native Americans were systematically degraded through national and local political and economic strategies. On the other hand, the harsh 'mental uniformization' in mission-led boarding schools attacked cultural integrity in order to bring their children to so-called 'progress' by enforcing another worldview on them. I dwell a little on this particular case since this museum allows for a voice by Native Americans from the north and the south, and thus is able to render a rather general or representative view on the history, detailing what happened over the last five centuries to these peoples in terms of 'development'. It is significant to learn that only on the occasion of the planning of festivities for five hundred years of Christian civilization (1492–1992) did Latin American critical historians and anthropologists finally manage to air an overview of the 'other' history; today there is a scientific debate going on about the 'genocide' of Native Americans throughout these five centuries. The debate has 'minimalist' and 'maximalist' positions: the former claims that some 74 million Native Americans were actively or passively killed by the civilizers; the latter holds that 94 million were victims of this 'progress'. Moreover, the estimate of the amount of Natives who were forced into slavery in the first centuries of white conquest varies from 2.5 to 4 million (Fisher 2017).

In other continents, other types of pillage or one-sided treatment can be pointed to. This leads presently, almost a century after the first attempts at decolonization started (after the First World War), to reports about the crimes committed by Dutch bosses in Indonesia or Suriname (Wekker 2018), Belgian colonial rule in Congo (Faassen and Verdijk 2017) and many others. Simultaneously, there is a great deal of discussion going on right now about 'cultural appropriation' in the past and the present, with a hefty debate on the restitution of 'stolen' ethnographic material that has been displayed in western ethnographic museums for over a century now (see, e.g., van Beurden 2017).

Finally, it becomes clear today that, on top of the colonial past, the expansion of the West in contemporary globalization is driving the very same formerly colonized populations towards horrendous survival conditions; indeed, the first victims of global climate change, of the destruction of biodiversity and of the new economic and social inequality will be the peoples of the 'poor south'. Islands in the Pacific are threatened with floods, parts of India and Africa are drying out quickly and will soon be uninhabitable for humans because of periods of extreme heat and droughts. I could also mention the effects of the present and future pandemics, triggered by assaults on biodiversity, where it is already clear now that these same poor peoples will be treated as second-rate patients when it comes to the availability of vaccines. A top researcher at the IPCC emphasized that even if technological innovations alleviate some of the negative effects of climate change, the former colonial populations will undoubtedly not be the primary beneficiaries of them (van Yperseele, Libaert and Lamote 2018).

This type of data, available and sometimes even recently rediscovered (see Graeber and Wengrow 2021), invites the reader to consider the humanist legacy from a different angle: notwithstanding very few exceptions, humanists in general did not really question the Eurocentric worldview which was, and remained, cast in the mould of the religions of the book. Even – or maybe even more so, if we take Graeber and Wengrow's critical analysis seriously – since the Enlightenment such influential thinkers as Rousseau (with his 'noble savage' concept) and Smith (with his emphasis on private property as untouchable value) gave a new respectability to the Eurocentric, exclusivist point of view. So-called freedom of thought and logical reasoning, as its main instruments, took the Mediterranean view on things to be sacrosanct, classifying what was foreign or 'other-cultural' as less human or at the least less civilized. Moreover, what was rediscovered from Greek and Roman thinking traditions was systematically thought and defended through 'Christian glasses' until the second half of the twentieth century (Veyne 1989). It is a bitter truth, but it has to be recognized in all humility, that humanism thus far did not escape from a narrow-minded Eurocentric frame of reference. It is my conviction, as a devoted humanist–anthropologist, that this parochial frame should be rejected by humanists today even if it is only because global interdependence forces

62 • Humanism Revisited

us to do so. An open-minded and truly inclusive humanism can then be developed, recognizing other perspectives on reality and on humanity as a basis for horizontal egalitarian knowledge and understanding of humans in all formats and traditions, and on the grounds of such understanding can look for genuine comparative knowledge of what is human and for commonly shared action plans in the dangerous predicament we have to live through from here on.

The Fundamental Choice for NE-Humanism

We can of course deliberately choose to stick to a Eurocentric view, in its *racist* or its *cultural identity* form. Several missionary programmes continue to invest in such an attitude: the bitter struggles on that front between different denominations of Christianity and of Islam in African, Asian and Latin American countries (but also in North America, as mentioned, or in the European countries – I think of Ordo Iuris and other movements on the internet nowadays: Höhne and Meireis 2021) tell that story. But profane programmes such as the educational programme of the OECD or of some large foundations also take that sort of approach on fellow human beings and their cultural traditions. I have worked for more than three decades in the field of 'ethnomathematics', trying to devise means and procedures that would make mathematical education emancipatory and powerful for populations living within a different cultural background, instead of driving large quantities of their children in the trap of what is generally called nowadays the realm of 'dropouts'. With a group of mathematics pedagogues and social scientists we try to combat this unifying and indeed imperialistic way of conceiving of schools as instruments to 'reprogramme' children of other traditions in the one, 'right' worldview of western thinking. Some of my colleagues would speak of continuous 'epistemicides' unfolding (Chronaki 2011; Chronaki and Lazaridou 2023). On the one hand some commercially interesting discoveries – such as, for example, the boat structure of the catamaran – have been stolen or 'culturally appropriated' in the course of history (Rubinstein 2004). On the other hand ways of using mathematical concepts and procedures from a non-market view of one's lifeworld (in ritual, music, cosmology, etc.: Pinxten 2016) are de-learned or dissuaded in schools, since they are considered to be 'the handicap of tradition' rather than an addition to western perspectives. The permanent exhibition about the missionary and state education of Native Americans in the Museum of the American Indian of Washington, DC is awkwardly telling in that respect.

Recently, the critical economic writings of scholars such as Thomas Piketty (2014 and 2019) or John Hickel (2021) have documented how growing inequality boosts this colonial attitude in the dominant society's educational politics of today. Even such major derailments as the climate crisis or the banking crisis of 2008–9 do not invite westerners to take a more modest, and indeed honest, inclusive attitude: for the moment I did not come across any programme in western countries (or in China for that matter) which seriously questions this monocultural, dominant exclusivism and allows for a deep, pluralistic approach, for example. Instead, when the peak of the crisis is believed to have passed, pedagogical voices emphasize that the time is ripe to 'go back to normal', albeit in an even sharper and more excruciatingly competitive modus: views begin to be heard, for instance, that 'fugitives' might be acceptable in Europe provided they fit into the rat-race model that is considered to be 'normal' and can easily be taught what is needed to fit in. The welcome of Ukrainian refugees into western European states is explicitly stated in such terms as I write these lines (May 2022).

In this way Eurocentrism stays intact or even becomes more dominant in the light of the emergence of new hegemonic competitors such as China. The latter, as we learn from recent studies, does itself not operate within a more inclusive or panhuman humanism, but develops ever-more sophisticated means and rules to make dependence on the one unquestionable central authority (versed in Han cultural terms) the exclusive way for humans. That China has been exporting this attitude to other parts of the world is now being increasingly documented: the testimonies of Hong Kong, and also of Uighur, groups within the Chinese sphere of influence are to be added to the many studies on African, European and other regions which are subjected to the doctrine of the new Silk Road in this era (Frankopan 2018).

Of course, all this is of relatively recent date. In that respect, however, it is important to remember that most theories, as well as practically all experimental data and models which are used to justify this loyalty to Eurocentric (or western-centric) views on humanity and its education, are heavily cast in many years of (implicit or naïve) exclusivist scientific work. That is to say, it is good to be conscious of the biases of the data and the models (including the quantitative approaches in psychology and some of the social sciences), especially those studies that are considered to be the 'more scientific' ones because of their quantitative approach. Earlier, comparative psychologist Michael Cole (1996) detailed how practically all these experiments used university and college students as subjects, since they were and are readily available to the university-based researcher. This cohort is, obviously, a particular segment of the population: they are overwhelmingly middle class, White and well drilled in the 'schoolish way of thinking'. Or, to put it in the words a Navajo informant used when I asked him about such measuring devices, to which his people were subjected on certain occasions (to qualify for a job, to get a grant for further studies, etc.), 'You know, we can do these tests (if they come to check on us), but we do not think that way'. Apparently, no educator ever asked my informant or anybody else from his people to explain just how they think and why they feel this is adequate for life.

In all fairness, humanism and the Enlightenment at least started on a journey of deep soul searching and honest philosophical reflection. Maybe, this critical and 'free' search in itself was and is the most important feature of the original movement. Still, the often-cited and more or less 'vintage' contemporary authors in this movement from Anglo-Saxon (or French, Scandinavian...) origins regularly send out messages that testify to another mentality. I am thinking of highly respected authors such as Richard Dawkins, Daniel Dennett or Steven Pinker. They are critical, yes, but that attitude seems to stop where other cultural traditions start. I here examine the last-named's most recent book, which carries a highly positive quote on the cover by Bill Gates.

In that book Steven Pinker (2021) gives an overview of the tremendous positive points of humanism and the Enlightenment, and situates the opponents of this shift in mentality in a camp of unwilling and/or erring conservatives. Pinker correctly starts his book with an enumeration of Enlightenment principles: reason, and therefore 'cancelling' the anthropomorph God; science as the sole basis for knowledge; humanism; progress in a market system; and peace. However, when he develops these principles in the course of the book, he disqualifies criticism by others first of all as a sign of 'progressophobia' (Pinker 2021: chapter 4). That is to say, according to him critics are driven by a glorification of the past and a deep pessimism vis-à-vis the future. This yields the denial of obvious progress in a series of domains, he says: Pinker lists statistical studies on the growth of life expectancy, health, sustenance, equality, environmental care and peace, for example. Having given the numbers he can then go on to the bigger and less quantifiable forms of progress in the domains of democracy, quality of life and happiness. The final chapter on humanism makes the point that, against the background of all these forms of progress, humanism (as in the Humanistic Manifesto III) is the only viable expression of the will to do good for all of humanity.

As a devoted humanist and scientific researcher, I feel deeply offended by the spirit (sic) which permeates the book. I wanted to have self-declared humanist thinkers such as Pinker as my allies, as fellow-humanists. Still, his views and the style of argumentation leave practically no room for critical examination itself, which is – at least in my view – a prerequisite for any humanistic perspective on life. Pinker's perspective is too partisan and closed minded for my taste. I list a few arguments here which will show why his sort of justification-thinking is not enough, and indeed blocks a genuine humanistic view on the world and on human actions in it. I list three types of critique:

1. *Quantification as such*: throughout this book (and similar ones by other proponents) statistic data are used in order to refute the claims of followers of religions, but also of critical voices in the secular camp. Over the course of time, it appears to me, there did indeed emerge 'camps' on the life stance or existential positions in the West. In Pinker's introductory chapters the focus immediately calls for a division into opposing camps: those who seek some foundation for the meaning of life in an open way, not predefined by religion or by secular organizational thinking, are accused of 'progressophobia', and often seen as unwilling to grant the Enlightenment the benefits it brought forward. This is done primarily in terms of a battle between ideas: thinkers of the humanist–Enlightenment branch are discussed in a broad 'history of ideas' where they come to oppose the defenders of Church and dogma, so to speak. I do not deny that there is some power of ideas at play, but I feel dissatisfied as a social scientist that this reference to ideas would be sufficient ground to stand on. In such a view on things it looks as if ideas as such set things in motion and steer the course of history. That is both pure idealism in a philosophical sense and a gross oversimplification of the way history is shaped, at least in my view as a social scientist. To put it quite bluntly: human beings do not live in their minds only; neither do they make life-stance choices on the sole basis of rationality. In the third point (with Graeber and Wengrow) I will detail where that led us in the case of the Enlightenment.

Indeed, a mere history of (philosophical) ideas is deeply insufficient for a social scientist, since it systematically excludes the impact of socio-economic and political context on choices and on thinking in general. Science (and technology) as the better example of rational searching does not develop in a void: some concepts and research programmes are promoted against others, because they serve the need of the political–economic contexts. Paradigm shifts happen not out of rationality alone but because the proponents of the rival perspectives die out (as Max Planck significantly stated, and Thomas Kuhn further elaborated) or because military and economic demands steer research in another direction. For example, just in recent years, medicinal research (such as the development of penicillin, but also of various vaccines) and also ICT developments have been promoted very heavily by economically and ideologically driven policy programmes, both in western democracies and in the USSR and China. The push for monetarist economic models, but also for research that produces patents, became a dominant dynamic in the past few decades of neoliberalism, where universities and research centres have been run more and more by business managers instead of scientists. In this shift, the emphasis is now on competition and market values rather than 'rationality' as Pinker would have it: this is what Sahlins appropriately called the 'corporatization of academia' (Sahlins 2008). Of course, when the inspiration, the basic concepts of research questions and the whole culture of research (in science, but undeniably also in technology) are thus to a large extent dependent on contextual elements like market and management logic, profit and so on, it is a little strange to uncritically see these contextual aspects as simply unimportant or even as neutral parts of the engagement for reason. If one can still speak of 'rationality' here, it is at best a narrowed-down version of it: short term, market-conforming. Concretely, when the statistical studies Pinker cites to make his point are meant to work as incontrovertible evidence of the 'human progress' he wants to illustrate by means of them, he omits to take these contextual elements on board as steering elements in the development of a humanistic and enlightened progression. That is, to say the least, very hard to stomach for a humanist like myself (and I guess for many European continental humanists): is it 'reasonable' or 'rational' to have scientific progress mainly as a by-product of war research in the nuclear sciences, or even IT developments in military circles, or the enormous pillage of natural and human resources in the capitalist development one calls globalization? On top of that, one should mention (as an anthropologist) the imperialistic policies of installing school education and its fixed (progressive?) curricula all over the world at the expense of oral traditions and of other inroads to rational interaction with the natural environment (see, e.g., Vandendriessche and Pinxten 2023).

Yes of course, disregarding all this and measuring to what extent the market models, as the sole expressions of 'progress', are scoring everywhere, statistics can be used to underline this sort of argument. But is this telling the story of more humanism made real, or rather an uncritical account of the success of one political–economic model without counting the side-effects, let alone the loss of human dignity and creativity, along the way? Statistical results show what they do in terms of the question you ask. The discussion on the choices of appropriate questions, on the concepts and ideals embedded in them, is not gone into by Pinker. For me, this sort of approach is hence lacking in respect and even in critical power. I would explicitly refrain from 'moral' judgements, since the more basic issues of a contextual nature are quite systematically neglected here.

2. A second critique turns on other blind spots in such treatments as Pinker's: anybody involved with the *economic studies* of late has to recognize that the progress Pinker is hinting at (see his chapters on inequality, quality of life and suchlike) will point up that over the past three decades inequality has been sharply rising (see Piketty's painstakingly detailed historical overviews: 2014 and 2019). After the 'heureuse trentième' (1950–80, a heavily fought-over compensation to the lower classes for the war efforts) which installed a more fairly redistributing political–economic policy, the rise of neoliberalism since the Reagan–Thatcher governments saw a soaring of inequality in the rich West and elsewhere. I cannot go into the statistics here, but ample researchers have been documenting this (Atkinson, Stiglitz, Piketty, Raworth, Hickel and so on; see also the NIC report of 2021).

The rise of 'bullshit jobs' (Graeber 2015); the financialization of the economy, with its new billionaires on the one hand and its 2 Euro/dollar jobs on the other hand; and the rapid growth of poor and ill-protected workers (especially in the Amazon and bol.com bracket) all testify to tremendous inequality gaps. In my view no humanist or Enlightenment defender can ignore this sort of development and keep on saying that it is progress and human happiness everywhere. Nevertheless, this is what Pinker and some other authors tend to do, because they base their analysis exclusively on a history of ideas; it is more difficult, but also more correct, to critically weigh statistical analyses and to base one's choices and evaluations on a deep and honest discussion about values and preferences in the contexts that prevail. The questions to be used for evaluations and broad-scale testing will have to be checked against this broad background, and in that respect statistics will then again be relevant dependent on the questions asked. The argument will not be found in the numbers as such, but in the numbers on humanly relevant questions.

3. In the third place, a critical analysis of the *mythology of progress* is in order. This is a difficult and indeed painful point. Over the past two decades several of the basic standpoints on western civilization and the 'others' have come under attack. This now proves not just a question of broad- or narrow-mindedness, but really an issue of fact versus self-glorifying myth. I explain this a little bit, and refer the reader to further sources for deeper analysis.

In the social sciences and in the political discourse of the European colonial powers (and, later on, of the West in general) the conviction took hold that 'we' developed nations possessed structures of freedom, democracy and equality, whereas the peoples Europeans had conquered since the time of Columbus had nothing noteworthy like this. This sort of conviction was especially fed by the way Europeans constructed a perspective on the peoples they encountered in the Americas, Africa and the Pacific area. Power and greed doubtlessly explain *why* this perspective became dominant. Of course, as an anthropologist working with such a deeply different language as Diné (the Navajo Athabaskan language), I can see *how* false knowledge and 'framing' in the colonial sphere of interest became likely in these cases. To name just two elements that explain this little history, there is the difficulty of translation and there is the lack of serious comparative methodologies until recently. The many different cultural traditions (the Human Relations Area Files system in anthropology reckons with up to four thousand living cultures) and the numerous and diverse languages (here the estimate goes up to six thousand genuine languages in the world, even today) show how the engagement to understand these 'others' sincerely and with enough depth is enormously taxing. But again, let us not forget that the Europeans went out to conquer – find gold and silver primarily, and many other types of natural wealth later on – and that they justified this by the solemn duty to try and convert the 'other' humans to the only right worldview, laid down in the Scriptures on the basis of the word by the Almighty himself. In that perspective on humans and on the earth (called Creation by that tradition) there was no room for broad interpretations, let alone for what is today becoming known as 'pluriversality'. Especially in the Europe of the fifteenth and sixteenth centuries, wrecked by endless religious wars, there was little room for leniency or open-mindedness. As we know, the early humanists were the first to learn about this structural intolerance.

A few centuries later, with the French Enlightenment thinker Jean-Jacques Rousseau and the English political thinker Thomas Hobbes – as well as with the first modern economist, Adam Smith – the frame of reference was installed that 'they' could and did not have history, and hence had no democracy either. Theologically speaking 'they' still lived in the dark ages of pre-Christianity. The recent book by Graeber and Wengrow (2021) offers a painstaking and very thoroughly researched analysis of how this picture became dominant. As the authors show, there was a time during the first centuries (especially the sixteenth and even the seventeenth century) when the voices of Native North Americans were heard in France and to a certain extent in England, but they were forgotten or suppressed later on when the colonial exploitation of the new territories came into full swing. But the first and more nuanced voices were heard and, to a degree, offered a platform in Europe through travellers, who indeed learned the languages to some extent and lived for decades with the 'primitives'. They learned of plural Nation Confederacies, of the native refusal to amass wealth, of the latter's reaction to the strange habits of the so-called civilized conquerors, of their amazement about the characters of God and the Holy People, and so on. Graeber and Wengrow relate how first Montaigne and then some of his fellow humanists saw and heard representatives of these 'savage peoples', and how Rousseau and his contemporaries heard them or came into contact with travellers – authors who told stories about them. Graeber and Wengrow are well aware of the ways image building about 'the other' was manipulated during the period of early globalizing capitalism, and they offer fact upon historical evidence to illuminate this issue. The whole frame of reference for the us–them distinction of the past few centuries is shaking: it proves to be that the first encounters held some quite substantial testimonies by Native Americans (and other non-western subjects) about lengthy, democratic decision-making procedures in their societies, as well as about deliberate and conscious mechanisms to reject or eventually surpass abuse of power – even the mere concentration of power in hierarchical structures. Again, it is significant to learn about the discussions going on between missionaries, settlers and Native Americans on these issues, stretching from the first encounters of the sixteenth century right up to the beginnings of the industrial capitalist era, which really took off together with Enlightenment thinking. The authors analyse how Rousseau's ideas on the 'noble savage' were first received as a conservative philosophy (much in line with the Christian view of us and them), to be turned into a policy statement that Europe (or later, the West) had the unique duty to 'liberate' these peoples by forcing them into the world of modernity through conversion and schooling. Again, I like to remind the reader at this point of the statements and the factual documents in the Museum of the American Indian in Washington, DC, which were reacted to with disbelief by many 'White' visitors I encountered. The Native Americans underline in the permanent collection of that museum the very line of historical data that was neglected for centuries later on. What transpires in Graeber and Wengrow is, one might say, a more systematic historical analysis of the Native American statements to be found in the museum. The implications of such a systematic and historically painstaking analysis are considerable. No, democracy was not uniquely 'invented' by Europeans. Yes, the us–them distinction in terms of primitive–civilized oppositions was framed against historical evidence to a large extent, and one needs to understand how political and economic interests may have induced this framing. And finally, no, the Enlightenment philosophy of Rousseau and Hobbes is not one of freedom and progress as such, but proves to have been used to favour the freedom and wealth of some at the

70 • Humanism Revisited

expense of many others. Along the way, the image of these others has been formed and transformed to a large extent to allow for these one-sided gains and profits of some of 'us' at the expense of 'them' – and, in truth, of a lot of 'us' as well. Again, as a humanist and as a scientist (and hence loyal to the Enlightenment ideals) I have to take all these contextual elements and their impact on the actors into account if I want to underline the possible benefits of the humanistic and Enlightenment legacy. If things failed, were abused or got derailed under way in any sense, I have to recognize that and start to work at a reset of the programme. Accusing critical voices of 'progressophobia' will simply not do, and might even qualify as dishonest.

But the situation is far worse than a mere misrepresentation of fellow human beings who became entangled in perverse processes. As I stressed in former parts of this book, the net result of the 'free market ideology', with free agents in the Hobbesian style working in a minimally structured societal or state context in Smith's view on the market economy is leading today both to a deep and dangerous disruption of nature (see climate change and destruction of biodiversity) and of equal opportunities in the world (with steeply growing inequality in recent decades). Here again, the tenets of the old humanism and of a rather exclusive interpretation of the Enlightenment (in the Anglo-Saxon view) need rethinking for the old ideals to be defended in an honest way.

What NE-Humanism Amounts to

In the light of the critical points I listed in the former section I want to invite all humanists, believers and non-believers, to engage themselves in a deep, critical assessment of several of the foundational standpoints of humanism (and the Enlightenment). The dramatic changes in the context that we, westerners, are living in today invite us to become more self-critical and maybe, first of all, more modest about the superiority of the models and values we have been promoting. In this chapter the first implicit premise is scrutinized: the self-evident value of a Eurocentric point of view. We should develop ideas on how and in what form a non-Eurocentric humanism could be thought out. In essence I see two main foci here, but the discussion is wide open and I have no intention whatsoever of presenting these foci as the ultimate programme for self-critical research.

Towards a Differentiation of Ethical Notions

As mentioned in the chapter on the humanistic legacy (chapter 1), the emphasis from the beginning was almost exclusively on individual rights and duties – and hence the power of individual choices and decisions. In the institutions that emanated from the humanistic tradition a lot has been said and printed within that individualistic perspective on things. For instance, in a very modest and hesitant publication by the national headquarters of the Humanistisch Verbond (Humanistic Union) of Flanders, Belgium of 1987, rather voluntaristic statements are found stressing that one should be careful not to overrate individualism, meanwhile disregarding ecological concerns. But unfortunately, the call for a self-critical reassessment stopped there. More than thirty years later little progress had been made on this point; consciousness of the Eurocentrism displayed by all these individual humanists, on the other hand, was never seriously discussed. Instead, with the growth of Islamic political actions in a context of growing 'oil-dependency' in the West humanists and their institutional organizations were most of the time speaking out from a fairly Eurocentric perspective about 'others': I witnessed more than a decade of bitter fights against the headscarf, which was unilaterally seen as a way to oppress women's individual choices. Also, I saw – in my professional engagements and in my actions as humanist and critic of discrimination – most often a unilateral enforcement of secular rules in the public sphere: 'no religious signs in any public office' became a fiercely fought-over issue. In the zeal put into this 'fight for individual rights' humanists became rather blinded to the fact that centre-right to extreme-right parties and pressure groups have been actively abusing this unholy emphasis on the individual's exclusive rights and privileges in order to spread fear of 'the other' and thereby broaden their political power. For instance, the 'cultural shift' strengthened by Huntington's (1996) book constructed 'us' as free and well-meaning individualists against 'them', the narrowminded and mostly collectivist subjects who were on the verge of attacking this western superior position. In Europe this shift has resulted in the growth in power of extreme-right parties, who presumably cleverly claimed they and they alone defended 'individual freedom', often together with a more or less humanistic set of values: the Front National in France, the two or three 'freedom' parties in the Netherlands, the Vlaams Belang (Flemish Interest) in Belgium and so on were the forerunners of the PiS (Law and Justice party) of Poland and Victor Orbán's Hungarian Fidesz party today. Some are secular, some Christian and some mixed, but all will defend 'our' rights in terms of the superiority of exclusive individual choice along the Eurocentric perspective.

In COVID times, since 2019, this rightist movement is reinforced by influential rumours and ideological models that spread the idea that even administrations are more and more guilty of deceiving the citizen in his individual decision-making power. Crowds in the streets shout in all our cities that 'our individual freedom' is under siege from the governmental rules that should enable our societies to counter the pandemic. Thus, humanistic 'individual freedom' is used – rightly or wrongly – and captured rather successfully by rightist anti-democratic groups (Höhne and Meireis 2021; Ponsaers 2021).

In a broader sense, along with Piketty (2014 and 2019) and Stiglitz (2019) I can only note that the supremacy of the material interests of a small segment of the world's population (i.e. the western well-to-do citizen, together with elites elsewhere) has not waned. As analysts and scientists keep on telling us in despair, we have known for half a century now (the reference point is the publication of the 'Report of the Club of Rome' in 1972) that we are causing the extremely dangerous climate crisis by our 'freedom of choice' market system, wherein humanistic principles may not be the main drivers (greed and egotism look likelier candidates) but they certainly have a major role in justifying the continuing pillage of the planet and the crazy, blind race towards climate warming. Since the debate in the West keeps the focus on the supremacy of the values of individual choice and the exclusive role of individual prerogatives the voices of others, thinking and speaking from different perspectives on humanity–nature relationships, are disqualified as inferior; primitive; or, at best, romantic remnants.

All in all, humanist organizations have reacted slowly and with tremendous reticence to this development. I suspect the general feeling is that 'politics' should be avoided, since engagement on politically salient issues foreshadows tensions and possibly splits within the ranks. For example, a 1987 publication by the Flemish organization of humanists (HV) was the first and, in fact, only attempt ever in the Low Countries to at least put the ecological issue on the agenda. However, the way it was done was telling: the authors recognized there was a growing problem, but in the end and in consensus they left it up to the voluntarism of each individual humanist to take the issue to heart or not. My experience, as former chair of this large organization and informed on stands of sister organizations in the West, was and is that indeed the main argument used in order not to have to act as an organization is that humanism is best restricted to the private or individual level so as to reach people across a broad political arena. That is to say, people from the centre, conservatives and leftists should be reached. By narrowing life stance to individuals and to voluntarism on the part of individual human beings, one avoids entering so-called political territory: individualism would keep different convictions within the one flock. Moreover, with the present-day distrust vis-à-vis party politics all over the West (Albright 2018) this attitude leads to a self-inflicted irresponsibility for humanists and their organizations: if you cannot have a clear position on societal effects of a rogue or pillage capitalism which is causing inhuman situations under your very nose you give away your presumed intellectual or conscientious power. You become guilty by abstention, so to speak. Put differently, it has become clear to everybody that nobody with a conscience can look away, but your choice of voluntaristic individualism continues to make collective reactions almost impossible for humanists.

Having said all this, I refer back to the short analysis of identity processes, split out over three levels: the individual agent (re)acting on her own, the group as source and target of action and the community as a set of virtually interacting units. It might be significant, I feel, that Hannah Arendt has (once more) returned to the heart of debates today. The evolution in her thinking about the possible further 'humanization' of the world is illustrative of the reactions vis-à-vis the idea of NE-humanism. During an initial period in her thinking she chose an existentialist, individualistic approach. She was subsequently rejected as a 'member of the club' by her fellow travellers, who all had – predictably – different opinions on fate, humanity and suchlike. Arendt expressed her disbelief over the idea that individuals alone would be the main source in a possible growth towards a more humanly responsible world citizenship. She started calling this 'naïve humanism' (Goorden 2019: 77). She thus came to recognize later in life the second and, moreover, the third level of complexity I mentioned. For the latter point, humans are part of and are co-determined by communities, which act and react as a whole. Her shift is telling when one appreciates the nationalist and increasingly antisemitic context in Germany at the time. In her important studies on totalitarianism and in her Eichmann book this emphasis is striking: she depicts Adolf Eichmann not as an individualistic, sadistic maker of choices but primarily as the uninteresting coward who grew into the 'effective' monster he was precisely due to the features of the community from which he emanated. The 'defence' Eichmann gave and even his obvious surprise about the accusations voiced against him offer arguments for that view. As I read Arendt, she never says that the individual has no say at all but rather that the larger, societal context does or does not allow for the ideals the humanists think are primarily within the range of the individual agent. Of course, this does not deny that individuals can make a difference, but it recognizes the importance of communities as agents as well.

When discussing the relevance and indeed the need for NE-humanism, I claim it is important for the humanist to self-critically reflect on his engagement by including the community level. Indeed, in my opinion as a social scientist, it is undeniable that deep and possibly unalterable changes take place at this third level of human (inter)action which cannot pass unnoticed to the western humanist.

74 • Humanism Revisited

In the first place I want to discuss here how interculturality has been becoming very prominent for decades now. It is clear that people are travelling massively all over the world (as tourists, as workers, as refugees). A net result of this is that the big cities and urban areas have become deeply culturally mixed over the past few decades. Concretely, the most urbanized part of western Europe – the rectangle between London, Amsterdam, Frankfurt am Main and Paris – has inhabitants from more than a hundred non-European countries and cultures today: all the urban areas have EU citizens of different countries, but apart from them they count numerous other cultural origins amongst their population. For example, a small city like Ghent, Belgium with some 260,000 inhabitants counts citizens from 157 different non-European cultural traditions today; Brussels has some 200 different non-European cultural origins in its citizenry; and Amsterdam tops the list so far, with over 220 of them. This means that food, clothes, language and views on gender and on child–adult types of relationships – but also on the roles of schools, social services, marriage and so on – are diversely conceived by many of these groups and communities, sometimes overlapping and sometimes conflicting with the 'native' community. The englobing 'community' of Ghent, Brussels or Amsterdam citizenry of course changes with the rather sudden and thorough mixing of their populations. Any humanist who is now confronted in her groups with a rapidly changing community should somehow learn to deal with this large and mixed entity of common agency. To state the obvious, the emphasis on the individual which was formerly seen as 'natural' by the Eurocentric humanist is not necessarily shared by many persons and groups from other cultural origins. How to cope with this? Should the humanist become a missionary and 'convert' those others to individualism, and on what ground? Or should this humanist look for different approaches, maybe differentiated according to the person or group she interacts with, and then how and to what extent can a common ground for 'societal' or community behaviour be sought for? Anyway, presupposing that 'all humans are individuals' is a rather futile and not very effective position in this regard. For example, what if clan membership is really very important and maybe even totemic religious activities over the clan (as I witnessed in my interactions with Ghent 'new citizens' from Ghana): how should this humanist deal with this, as an individual? But could she then avoid thinking and acting on such questions on a community level?

At the same time, and in the light of the present disruptive effects of climate and ecological imbalances on a worldwide scale, westerners are rediscovering economically and socially decentralized 'commons' in the wake of recent economic crises (Bauwens 2013). Apparently, more often than not such initiatives in the West are inspired by traditional, native/

non-western knowledge and action formats from elsewhere, for example from South American Indigenous peoples (Lotens 2019; De Munter 2010). Others will find inspiration in eastern cultural traditions: the growth of spiritual, therapeutic and also food styles from eastern vintage in the West has been growing steadily over the past half century. In a sense, reasoning along the critical lines of Graeber and Wengrow (2021) one could consider the present 'import' of indigenous ideas and practices as, at the very least, a second cultural wave. The first wave was the one these authors tried to dig out of oblivion where it landed in the course of the European capitalist conquest of the earth (starting in the sixteenth, but coming to full effect in the nineteenth and twentieth centuries). If my hypothesis is reasonable, then the humanist or enlightened thinker of today should not make the same mistake as yesterday, simply denying that other cultural traditions (now close to home and appearing to stay around for generations) either do not touch on 'our culture' or should be erased as quickly as possible (by means of an oddity called 'assimilation', turning the foreigner into one of 'us').

Let me offer an empirical example, once more, showing how things develop in the real world, one step removed from the mere history of ideas. The example I relate here illustrates how individualism is as much the product of a particular cultural context as any other perspective. The indigenous researcher Marisol de la Cadena (2011) reports on the cultural–political life of Turpo, a Quechua shaman from Peru. She stresses on several occasions how Turpo 'is called' by his community (a village) to channel his gift into a political format and become a candidate for elections. When, later on, he is invited by the initiators of the Museum of the American Indian in Washington, DC to install the Quecha part of the exhibits (2004) he is inadvertently seen as a personality in whom the President of Peru shows an interest. When that president is seduced later on by a multinational corporation which aims at exploiting the land beneath Quechua territory, it becomes clear that the Quechua culture and ways of life are seen by the multinational enterprise as an obstacle to its plans for progress. De la Cadena goes on to show how Turpo is involved in the process as an individual but at the same time as an agent for his community, who is built up by and dependent for his meaning and decisions on the local community. Individualism in the western sense (the autonomous choice of an individual, as if an island on his own) does not pertain here. Turpo has a 'calling'. He was, so to speak, predestined to use his gift in order to serve his people. He decides to follow this calling, even against his own will. Like in the case of his father this entails that he will have to fear for his life from time to time, and also that he is manifestly navigating between external pressures and the values of his people in his

76 • Humanism Revisited

contacts with powerful brokers: the army and some Quechua of his own community. In the discussions de la Cadena studied, the president increasingly follows his own private interests over the course of time and starts making statements on progress and modernity where individualistic rights of the 'White' western tradition are invoked and shamanism is even represented in the 'modern' clothing of a market phenomenon like anything else. This market, the president claims, emanates from a minority group like the Quecha (numbering some 400,000). When Turpo keeps refusing the multinational's deal the president changes his story and claims that the community rights of the Quechua are worth next to nothing in comparison with those of 35 million Peruvians overall, since the Quechua rights in fact belong to a former era. They are now said to block the 'universal' right to progress.

Apart from the corruption of the politicians involved (as became clear over time) I want to point to this case as an example of what is happening everywhere in the contemporary world.

At the same time, in attempts to overcome the mess and the life-threatening situation we are rushing into because of unlimited exploitation of the earth and people in the global economic conquest of late, this same 'primitive' or 'past' view of the relationships between humans and the rest of reality seems to inspire the new wave of commons, which is emerging in the West and elsewhere. The reference found in the case of the Quechua is to 'Earth Beings' – that is to say, all phenomena in nature are seen as part of, and hence linked to, each other in this one world of experience, in which humans are but one partner (see the discussion on ontology in chapter 3). The individualistic choice for private profit over and against the rights of the complex, integrated conglomerate of the 'Earth Beings' is alienating, according to Quechua understanding. In this example, as in many others related by anthropologists and development workers, the Eurocentric conceptualization of reality (including the value system involved) clashes with so-called traditional views; in the latter all phenomena are interconnected in a horizontal plane, one might say, and respect, consciousness of mutual impact and so on are the primary mechanisms and values, not individualistic views and private interests. I refer again to Descola (2005): when the western naturalistic separation between humans and nature is not subscribed to (which is the case for Turpo and the Quechua) and the holism approach to reality is adopted or safeguarded, then identities and ideas about agency on all three levels of action (individual, group, community) are shaped differently. The individual can never be thought as an island unto himself, and hence individualism as promoted by western people appears as an oddity – and maybe even an irresponsible and possibly dangerous constellation. This does not mean that people have no individual qualities: Turpo had a calling because of certain features that were particular for him. Of course, such features are only or primarily relevant in the context of the Quechua cultural community. Individualism in the western case conceives of the individual as a more or less independent entity, who is expected to think, choose and act on his own, regardless of the group or community context. In that perspective the individual has rights, which make him a sovereign socio-political agent. In Turpo's case, as in countless other non-western examples, this is an oddity, through which the interconnectedness with other phenomena is reduced or even denied. The mention of the Quechua case makes clear how Eurocentrism acts and what reactions it causes. This is, of course, just one particular example. In a useful overview Sandew Hari (2023) tries to sketch the total panorama of many thousands of non-Eurocentric initiatives worldwide. Awareness of this glut of alternatives is important; what I try to do in this book is to look at things 'at home', in the West.

Focusing on recent upsurges of individualism through that 'anthropological lens', the claim by anti-vaxers today that their individual freedom is being hampered by community regulations is beyond reason for all those 'others' anthropology is taking into account, while it is at most an exaggerated but recognizable consequence of individualism in the western construction of the individual. Conceptually and ideologically, this individualism grew out of the liberating reaction of the old humanists to the collectivist domination of the Christian Church. My reasoned suggestion with this book is to reframe the resulting, understandable emphasis on individual rights and preferences in all their exclusivity. This should be done with the knowledge we have now gained on the negative impact of so-called hyper-individualism in the capitalist era we are living today, within a context where 'others' should be finally seen and treated as fellow human beings and not as inferior specimens of humanity. The need to reset is even more urgent when we have seen new and more excruciating forms of domination appear over the last couple of years, where this individualism seems to be used in a perverse way: the growth of IT giants (Zuboff 2019: chapter 7).

Happily enough, this first road to reorientating humanism is open for the critical thinker: the many anthropological studies of non-western cultures, as well as greater access to ideas and concepts sent out to the West by the latter, can be listened to and discussed in an open-minded way by western subjects who are alarmed by the destructive effects of the aggressive and rogue capitalist globalization, often framed as instances of individual rights and choices.

The Political Dimension

When I qualify the critique of individualism primarily as an ethical issue, it is important to emphasize that more than individual–ethical points need to be made in endeavouring a reset of humanism. I will call this more englobing level the political dimension. Economically and ecologically the world we are living in today has become factually interdependent. Our political and ethical thinking has to follow suit. As one can witness with every crisis in international politics today (e.g. the sequence of COPs), this change in political thinking has not yet dawned in the minds of our present rulers. As I write, a military conflict between Russia and the West over the former's invasion of Ukraine is shaping up in the background, and for a couple of years a possible clash with China has regularly been reported on the front pages of newspapers. This testifies to the 'old view': states and the economic interests of particular, local conglomerates are said to be the main agents in the world. This amounts to denying – or rather, not recognizing – the global interdependence of all human beings. At the same time, it shows a blindness about human interdependence with 'natural' phenomena (such as climate, biodiversity and the like).

When discussing such matters one inexorably hits on deep-seated values such as limitless growth of the market economy, the right to patent almost anything (including air and water, but also human DNA: Mattei and Nader 2014), and so on. Of course, as mentioned before, some groups within this globalizing political economy – the 'commons' movement, repair shops, fair-trade organizations and suchlike – are trying to develop alternative perspectives, but so far they have been successfully framed by the dominant 'old' power circles as marginal and even as romantic remnants of the past. A next step, beyond mere 'soft' enterprise, is emerging, though. Economic anthropologist John Hickel (2021), for example, develops ideas on degrowth in the context of the crises we are living through, and political–economic journalist Kate Raworth (2017) has worked out a practical guide to escape from the dangerous road we are travelling now.

To be sure, such a shift towards a new political–economic model with the recognition of factual global interdependence involves very deep structural changes, and adjustments also in the minds of people. It is here that humanism has to get involved with political choices as well, I claim. For example, I cannot see how a substantial limitation of the right to private ownership, patent rights or the right to sell or simply dump (in the oceans or in poor countries) the waste of the rich and wealthy of the world can continue as 'sacred' principles once the notion of factual interdependence is deeply understood. This implies that community and global interests will have to be granted supreme status in several domains, at the expense of individualistic privileges. For one thing, this shift will imply serious soul searching on past colonial involvements (e.g. the genocide of Native Americans: Fisher 2017). The change is slowly setting in today, however, as such sensitive trends as the cultural repatriation movement for stolen 'colonial treasures' shows.

But the next step will be even more important: how are we going to avoid new forms of exploitation in the future by the powers that be, and plan out a type of updated, respectful, postcolonial humanism? What necessarily political stands should this humanism defend, whereby factual interdependence on a global scale is fully digested and the equal rights of peoples and of natural phenomena will be built in and effectively followed by individuals and governments? Such political choices will have to be discussed and decided on by humanists too, in contrast to the ways of the past where humanists essentially withdrew from the political dimension into the 'safe house' of the free individual. Whether we like it or not, political choices will have to be coloured by humanist values or humanism will become a marginal phenomenon of the past.

If I am right on this point, it implies that political issues should become part of the panorama of contemporary humanism. The derailment of hyper-individualism in the prevalent globalized capitalist regime leaves us no alternative, I claim. In the political context of today, with a reshuffling of power relations and maybe the emergence of new hegemons (China, Brazil?) and the decline of old ones (European countries, the USA?), a humanism-inspired ideological perspective is no superfluous luxury. Warnings about the death of democracy (Levitsky and Ziblatt 2018; Albright 2019) point to the urgency of resetting the humanist frame of mind.

Having said all that, what, then, would such a reset place on the agenda for the coming years? I cannot possibly offer a full-fledged alternative here: this will have to be discussed together with many well-meaning groups and communities. I can sketch a few points, though. One major issue I want to stress is that of cultural diversity: in contrast to the defensive attitude of combatting diversity in the name of the presumed universality of the western alternative, cast in a sort of ironclad set of values and perspectives over the past few centuries, the superiority of this universalism should be critically analysed. On this point, again, tentative moves can be perceived in the fight against different sorts of discrimination, including blatant racism.

With this fight, and in the search for another attitude, the value of solidarity should become a central issue, I think. Beyond the more or less voluntaristic *caritas* of Christian heritage, the fact of interdependence forces us to step beyond the self-satisfactory status of the 'freedom of choice' notion. Both the 'liberal' freethinkers (such as Pinker) and the European leftist humanists have been claiming that freedom of choice is a necessary, but also a sufficient, condition for humanism to persist. The final agent for them turns out to be the individual, since she is the seat of all choices and the 'measure of all things'. As shown by the histories of climate derailment and of surging inequality, the developments of the past three or four decades seem to question the validity of this tenet. There might be no other way than to situate and, to some extent, constrain individuals through group and community rules that reduce individual rights and subsume them under rules for humanity and even humanity+nature as supreme holders of rights. We might all have to choose respectful interconnectedness, yielding structural redistribution on a worldwide scale as a major value. Part of the discussion should then also be on the rights of non-human phenomena, including the climate, oceans, other species and all sorts of natural resources. Blatantly anti-humanist proposals such as that of former US President Donald Trump to 'buy Greenland' so Americans would be able to exploit the resources buried beneath this huge island should become impossible or perhaps even be labelled as criminal in a new, updated perspective on humanism. One of the lines of reasoning one can develop here is that of pluriversalism, thus going beyond the Eurocentric view on humans and nature in a fundamental and more daring way. Just considering this term puts a question mark on the old universalistic pretensions of knowledge, but certainly also on ethical values and concepts of law.

Of course, along the way it will be important to involve all communities in the world in such a trajectory: it is not only western groups and individuals who have been guilty of discrimination or inhuman treatment of others. But since humanism is first and foremost a tradition of the West, its overhaul will certainly be an issue there in the first instance.

These are a few elements I situate under the heading of an NE-humanism. In order to give the reader a greater appetite to start work along these lines, I now offer some concrete examples.

Some Examples of Eurocentric and Non-Eurocentric Thinking and Acting

Buffalo Bill was a retired military man, who shot many buffaloes (and probably also Native Americans) during his career in the course of the nineteenth century. It was the period of the annexation of western territory in the USA (the slogan 'Go West young man' summarizes what was in the air). Afterwards he had a second career as an entertainment tycoon: he toured throughout the western world with a kind of circus staging the cowboy version of what became very popular in the Hollywood film industry: how the

West was won according to the White folklore about it. This enterprise of Bill Cody must have been substantial at the time: he even came to Ghent, a small city in Flanders, Belgium at that point. After his show in Ghent a group of local sportsmen decided to start a soccer team (in the backyard of the textile factory really), which they christened 'the Buffalo's', thus honouring the entertainer. Of course, Bill Cody presented a romanticized version of the Native American in his shows, much along the lines of that other fiction-writing hero of those days, the German author Karl May. In both cases the Native American, played in Cody's show by the historical Sitting Bull on occasion, had to act as the primitive who was confronted with the choice either to be civilized by the superior Christian westerners or to die out. One can also think of the *zoo humain* format which became fashionable in the World Fairs in the nineteenth and twentieth centuries (Ben Chikha 2017). A few years ago this local soccer team (with international successes) was targeted by a human rights activist, who accused the club of making profits in a colonial style with the name and legacy of Native Americans. In the panic of the moment the board of the club invited me to reflect on this accusation. I explained that indeed the image of the 'Indian', which the club presents in its folders and promotion material, is colonial in style. I advised them to adapt the imagery and make it more respectful to Native Americans, and – being a rather well-to-do club today – they could maybe develop a sister club on one or more reservations in the USA to train young people in the noble sport of soccer. The discussion was lively, but the mind switch I was asking for was still too much for the puzzled club members. The basic reaction can be summarized as follows: the name and the emblem of the club is ours, and no foreigner (sic) can take them from us. I mention this small incident to illustrate what grassroots attitudes and 'knowledge' amount to. Even if one agrees that alienation is rampant (in a Marxian sense or otherwise), it at least stands out that the mentality of the 'common westerner' is not moving on towards a more inclusive and hence non-Eurocentric version. A better negotiator (who is more knowledgeable in the soccer world as well) might pick up such examples and bring them to successful alternative solutions.

A second anecdotal example refers to the policies of political parties in western Europe vis-à-vis cultural programming. Many cultural initiatives are dependent on subsidies, which involves the drafting and defence of dossiers with local policy makers to secure grants for cultural programming. In the wake of decolonization there was growing interest in cultural manifestations centred on artistic presentations from other cultural origins, and on the introduction of immigrants to cultural products of local vintage. In the early days after colonialism (the 1960s) there was a rather naïve approach on these issues: time after time, programmes were set up that

82 • Humanism Revisited

taught the 'others' how to learn to cook, listen to the music and eventually learn the crafts and the artistic tradition of the European society they had migrated to. During the 1990s a Flemish professor in philosophy, heading a ministerial committee, even devised a programme of x points which would be the minimal package for the migrant to know and appreciate if he wanted to settle in a European country. As often, this expressed a 'charitable' version of assimilation: we want you to become civilized, was the bottom line. Today, and especially after the Islamist attacks of the twenty-first century, this approach is dying out to be replaced by what I call a softer version: current ministers of culture decree that artistic initiatives should engage 'people from other origins as well', and should reach out to a culturally mixed audience so that everyone will be able to benefit from our great cultural tradition. With the coming to power of Trump in the USA and of a series of rightist politicians in all European countries, the latter aspect is stressed most of all: cultural initiatives should recognize the us–them distinction once more and eventually lead the 'other' to come and appreciate 'our' artistic products. Needless to say this approach is divisive and even polarizing in the population, yielding a sharper contrast between urban and definitely mixed populations and village or rural areas along the way. Again, these developments show that the fact of interdependence is not generally recognized in the political world, and that rightist identity politics is jumping into this crevice of ignorance and anxiety to solidify its power.

A Parenthesis: Philosophical Debates

Of course, courageous examples that run in the opposite direction can be pointed at. But the battle is on, and apparently 'culture' in the broad anthropological sense is a main focus: the culture wars that Huntington was calling for play a central role today, albeit that anthropologists are not invited to the debate by the protagonists of identity politics. Naturally, this is an old and continuing problem for western philosophy in general and humanism (and the Enlightenment) in particular. As mentioned before, the humanists of the sixteenth century were struggling with their Christian background, which could only understand human beings as a species created by God. Hence, discussions raged for centuries on the status of 'others' within the one and only Creation the European thinker was able to imagine.

I am not going to delve into the history of ideas in any detail in this book. Still, it is impossible to deny that ideas and philosophers have had an influence in Europe, adding on to that of theologians. The reason I will not go systematically into philosophical debates of humanist and Enlightenment thinkers over the years is that, as an anthropologist (with a philosophical

NE: Non-Eurocentric Humanism • 83

training) I am a disbeliever or maybe even a denier of the status of sources of historical change of philosophers in our, western history. I claim that the contexts these thinkers lived in is systematically forgotten or ill represented by philosophers: they have almost entirely focused on texts, on a history of (only) ideas. In anthropology (and many social sciences) the emphasis on understanding the broadest possible *context* of subjects makes all the difference for understanding why people think, value and do what is summed up as their particular culture. We are all capable and vulnerable human beings, and choose particular values and interpersonal relationships on the basis of many contextual elements that we are living with: climate, the food to be found, other species, ecology, history, social traditions and the psycho-physiological makeup of the human species. Within such a total frame one should understand what ideas and, eventually, what conserving or innovating projects are found. And the same applies to western philosophers: they do not create new realities, but they at best capture and further this or that view on things within the contextual conditions that are theirs.

Saying this may sound reasonable, until one meets theologians or philosophers. A great many of them tend to think that contextual elements need not be studied at all, since the ideas as such (the Creation story, or the power of detached human thinking) matter, and nothing else. When I grant room for the interaction between thinking and living contexts, the long history of textuality (emphasizing that we look at and live the world through texts: Said 1978) becomes debatable.

With regard to the main issue of this book this sort of argument is crucial, possibly because it seems so difficult or even blasphemous to hold. When one reads theologians, the shared perspective is that empirical aspects do not matter. If you formulate a problem, you should go and look for the answer in the holy texts because God supposedly 'expressed himself' in or through them. Hence, all relevant answers – and certainly the true ones – will be found there. Do not bother to study empirical data, but rather interpret lived reality through the lens of the holy texts.

What western philosophers have been doing is not so different from the theologians: the humanists and Enlightenment thinkers were and are critical of the theocentric stance, for sure. But while granting humans autonomy and denying the heteronomous power of the religious worldview they started a tradition that at least partially continued an attitude towards reality that can be identified in the theologians: in a sense, one can say that philosophers continue to claim that a detached, context-free type of thinking is possible, and may even be all that matters. Thus, they refer primarily to other thinkers and their texts. Humanists were awed by the rediscovered Greek ancestors in the field. Typically, they would reject sophists, who aimed at solving problems with their fellow citizens and in

84 • Humanism Revisited

recognition of the context they shared (Perelman and Olbrechts-Tyteca 1957; Toulmin 1990). Instead, philosophers would go for logical consistency and for concepts that appear to be beyond the level of the empirical or lived reality. Later philosophers then discussed and combatted the proposals of their forerunners in order to supplant them with new or slightly amended ones. The message read: empirical reality is not relevant, and logical consistency or coherence is the primary value.

When reading on cosmopolitanism, for instance, this remains the primary focus we find today: the work by Ralf Bodelier (2021) is a good illustration of what I want to point out here. This theologically trained philosopher works most of the time in Malawi, where he confronts poverty that we, westerners, can hardly grasp any more. He searches his soul in order to understand how western thinkers of past and present can help him to choose the right and morally noble reaction, being the well-to-do westerner that he is. He discusses five western thinkers with cosmopolitan proposals in his book, mentioning others along the way. Let me illustrate what I mean by my criticism of the attitude of textuality in philosophy by mentioning a few of his remarks on Kant. Bodelier rightly states that Kant's thinking has been an ominous point of reference for societal thinking in the past two centuries: his ideas on cosmopolitanism, on the role of legal rights within a sovereign state as the expression of the social contract of citizens of that state, and so on are taken as the standard way of thinking by contemporary influential thinkers such as John Rawls and many others. Reading histories of philosophy, it transpires as if the ideas thought out by Plato right up to Kant and his contemporaries are what matter most. However, Bodelier remarks that Kant, just like Hegel and more recent philosophers, reserved a separate category for those who did not follow this particular line of human history, the so-called Others, in a remarkably clumsy and often outright racist and exclusivist way. The last published book by Kant (his so-called anthropological studies, worked on for over thirty years) lists horrendous and racist ideas on women, Black and other people of colour. They are thought to be incapable of correct thinking and hence qualified as childish, basically irrelevant, co-humans. Bodelier (2021) rightly addresses this point, since he wants to use the treasure of philosophical ideas of the West to help him make the right choice in the poverty context he confronts daily. Hence, the author decides that we cannot accept this 'anthropological' book of Kant, with its racist and other exclusivist ideas, as a significant work: we should concentrate exclusively on the so-called main works instead. In my interpretation, this might save the philosophical ideas of Kant within the textual tradition but it certainly continues the blatantly exclusivist position of the western-biased theologians and philosophers who reasoned about value and truth only within the frame of the holy texts, disregarding all contextual elements. Acting on the basis of the authority of such textual ideas will be comfortable for the believer, but it most certainly is deeply imperialistic when other traditions (and, at the wider horizon, other species, the climate, and suchlike) are approached in a more respectful and open-minded way.

Presenting our western thinkers as superior per se and persisting in a tradition of textuality without serious attention to contextuality will inexorably lead to more, often sophisticated, forms of racism and other petty prejudices. The present rise of White-supremacist thinking, *Umvolkung* ideologies and the like testify to that effect. Moreover, the mere framing of one's choices in terms of law (as philosopher from Kant up to Rawls have been doing) does not allow us to go beyond this pettiness and mental imperialism: excessive juridical constructs and practices will emerge, and rights may be declared to be universal, but oppression and exclusion will not be vanquished by virtue of this declaration (Said 1978). A serious study of others' and of one's own contextual predicaments may at least yield an instrument for recognizing and evaluating premises that are tavailable.

This, then, is my invitation to humanists today: start thinking about your own Eurocentric prejudices or premises. In a second step we can then meet 'the other' (as in Emmanuel Levinas' work; this might even be a first step, since Levinas at least allows for the 'other' to have an important role and voice in the humanistic endeavour). I will come back on this second step later in the book. In a very succinct way, I see two possible roads here: the first one takes for granted that humanity can only survive if it sticks to one value perspective, reducing the 'other' to a derelict, a remnant of the past of the species. I see two versions of this choice in the present world, both excluding the 'other' in any deeper sense: one is the authoritarian top-down view in Chinese politics, colonizing or outright erasing difference (Frankopan 2018); apparently, the recent political evolution of Russia might be qualified as another version of this. The second road looks like a softer variety of the first one: it is the market ideology which has thus far spread by education and international law and proclaims that human beings are all market players and the so-called freedom of and within the market mechanisms will justify erasing 'the other' in the name of progress. Putting it like this may appear to the reader as rather crude, but it suffices for the moment to summarize it.

The alternative to both domineering traditions is one where we respectfully and in an egalitarian perspective picture the 'other' and ourselves as one species with a relatively deep diversity in the way humans, other species and reality in general are thought of and drawn into interaction. Here we find that the intrinsic dualism of the western perspective collides with the holism of several other human ways. When we self-critically manage to gain knowledge of ourselves and of 'others' we can then look for ways to negotiate with all and 'everything', in order to make choices that will be needed for the survival of the earth and of humanity. These choices will have to be made in unison, in stark contrast to the political choices of the two aforementioned political traditions. In my opinion this negotiated, open-minded approach is what could be called an updated humanism. It may take hold in either of the two imperialistic approaches, I am convinced, and this will obviously require giant steps.

Chapter 5

E

Ecological Humanism

The shaman spoke in the usual symbolic terms about the thirst for blood (or rather for life liquid) of the deceased, whose soul was said to roam the premises, invisible and with no foreseeable trajectory. The corpse of Constantin had not been handled in the proper way: the heart had not been punctured correctly by the young man in the village, a newcomer at the job. My friend Jean Bernabé, a rather sophisticated poet-turned-anthropologist stemming from a bourgeois family in Ghent, Belgium, was doing research on death and dying with the northernmost Romanian peasants. And then someone died in the village. As was usual, the local Orthodox priest had not been invited to the ceremony, but the communist district leader was there together with most villagers. Jean explained to me that in 1974 (the period of which I am speaking here) the balance between humans and nature was the cosmological frame for action, certainly on important moments in life like birth and death. The local cosmology explained to the peasants why the life-threatening disease of pneumonia (which is the name given by modern physicians, of course) returned from time to time and killed villagers: the soul of the deceased had to be set free in the proper way. If the latter failed to be the case, the 'one who comes back' (the vampire, in Hollywood terms) would return, in search of the life liquid: blood. Preparing the deceased in the proper way involved laying the body in the coffin with the eyes still open, giving it all that would be needed in the afterlife (food, clothes, a burning candle in one hand, etc.) and then closing the eyes of the dead person. After that the specialist comes in and pierces the heart with a wooden pin so that the life fluid runs out. If this last act is not done in the proper way the soul will come back

88 • Humanism Revisited

and haunt the community, since it is believed to be in search of the fluid: the blood. When all is done properly, then the priest is invited to prepare for the burial in the village church. After that, for forty consecutive days water is poured on the grave to create a 'river' by means of which the soul can make the passage to the afterlife. The priest is not invited to join in this final procedure either. Needless to say, several things could go wrong in this complex death-and-burial sequel (Bernabé 1981).

A few years later I had the privilege of working with the Diné speaker-philosopher Dagha Chi'lii. He gave me an intriguing allegory he had thought up himself; it relates to a derailment, a disharmony in the world of animals. Because the eagle was too greedy he started killing other animals out of sheer pleasure and a feeling of power over the rest. Death was everywhere. The other animals were on the verge of being slaughtered and might even have disappeared from the earth. Then they decided to collaborate with each other and teach the eagle a lesson, so he would respect harmony (*nizhoni*) again: they disappeared and hid in a large cave, unknown to the killer. The eagle went hungry, but he did not find anything to feed on any more. In the end, starved and weak, he found the cave. The other animals explained to the eagle that living in harmony is the only way to keep living in the long term. The eagle should respect that harmony as well. They reached an agreement and life could start again. Dagha Chi'lii gave this story as a present to me at a moment when I was about to return to the 'Anglo' (the American, White) world.

Do I believe in vampires, or in talking animals? Is there a need for such belief on my part – or, more importantly, is the need there on their part? No, these are instances of storytelling and/or ritual acts which express the relationships between phenomena in the world. And people learn about reality by means of stories first and foremost. The two stories I have related here deal with the proper, the harmonious or the sustainable relationships between existing beings. In that sense they are stories about ecological relations, in our modern terms. They deal with some aspects of harmony in the world, i.e. ecological balance, not by means of dry, cognitive analysis but rather by means of an allegorical tale.

We know the first humanists reacted to the corruption and abuse of power by the Christian institutions near the end of the Middle Ages, at the dawn of the Renaissance in Europe. The satirical tone of many of the humanist works, and their use of allegories, was meant to invite readers to 'deconstruct' (in our conceptual mindset of today) the complexity they were living and to try and distinguish truth from make-believe and power talk. In a predominantly oral context in Europe (with an estimate of more than 90 per cent so-called illiterates till late in the nineteenth century) storytelling was a highly respected and vastly used genre. Lately – that is, since schooling and hence literacy became the rule – the appreciation of oral forms has been gradually switched around. Instead of a genre to convey complex explanations about reality, it has been characterized as primitive and mostly 'false' knowledge about the world. At best, it could be used to entertain children. I have the idea that for the complexities we live with today, and which we have to explain in a language that is not buried in learned and sophisticated statistics and technical terminology, storytelling and the development of appropriate allegorical images and terms will be useful. Truthfulness isn't even an issue here: this is a genre, and hence has to be judged on its power to convey ideas; to offer material for argumentation; and to stretch the categories of running convictions, and not just as scientifically grounded statements of truth. One of the motors, or at the very least justificatory mindsets, for western culture today is found in the humanism (and the Enlightenment thinking) of some centuries ago. My point in this chapter is that the humanism we use as a frame of reference for life-stance choices and decisions has to expand and integrate ecological awareness. And maybe, along the way, produce new stories.

Ecological Thinking: What Is Meant Here?

But what would ecological humanism amount to? Let me return to the old and, at that time, revolutionary principle of the humanists: human beings as the 'measure' of things. Back then this meant that it was not God's rule (as laid down in the holy texts) about right and wrong, and about good and bad, that should be used as the sole and final criterion or 'measuring rod' to sort out which choices or decisions are allowed for, or maybe even dictated to, humanity. Rather, the choices and the moral principles developed by humans and considered good for them should come first. This may be a core principle in the powerful moral and political value that emerges: human dignity. In the minds of the humanists it is not God's will in the universe as He sees it that defines what this dignity has to come down to but the thinking and choice processes of human beings, taking their own insights and values as ultimate criteria. As everyone knows who is aware of the past few centuries of philosophy in Europe and the West, this notion of human dignity became a foundational principle for all political and legalistic philosophy since then (from Kant right up to Rawls and the Universal Declaration of Human Rights). Whatever practical arrangements were worked out for democratic rule in the past centuries, the principle that humans are masters of their own fate and that their choices and acts rest on human values and will be evaluated in the end on their effects for human purposes, was an undeniable tenet. We humans are the final judges, and our values (interests,

90 • Humanism Revisited

convictions, needs …) are the horizon against which we judge. God (in the sense of the Mediterranean God-Creator) was progressively pushed back out of the world of experience. He is still present in the American Constitution of 1776, where one reads that 'all humans are *created* equal'. But there, God the Creator is used as the final justification for colonization: neither Native Americans nor the imported African slaves were included in the document's notion of 'humans', and certainly not of citizens, until late in the twentieth century. This notion had already vanished by the time of the French Declaration of 1789: 'all humans are *born* equal'. In practice God disappears progressively and to different degrees from legal and political discussions and treatises on how the world should be organized, to be replaced by human choices and interests, taking the stand that human needs and insights are the rule. That is the humanistic meaning of the principle that 'man/woman is the measure of things'. Of course, American and Russian leaders will still invoke God as the final reference, where most European leaders will refrain from that (in the only genuinely secular area of the world), but, since modernity, the letter of agreements and of laws has not used religious convictions as an argument in most political and legal decrees in the West.

With the growing crises of climate warming and of biodiversity destruction, this rather exclusive focus on humans as unique and final decision makers is becoming problematic. My claim is that this context was missing from the original humanistic frame of reference, since the rebuttal of heteronomous power was its first aim. But the crises we humans have brought about over the past few centuries urge humanists to mend this blind spot. From the perspective of Renaissance and Enlightenment thinkers, the wealth of nature and the resources to be found in it for the realization of human projects must have seemed limitless, beyond possible exhaustion. It seems to have been the general conviction that whatever humans took, exploited or used from nature would be harmless or would be restored over time by nature itself. The famous capitalist dictum 'the sky is the limit' expresses this basic idea to this day. Responsibility was and is limited to the human world, first and foremost: an overwhelming amount of regulation and law deals with human benefits and/or misfortune; nature, by contrast is inert – it has no rights.

Of course, such reasoning and this kind of agency can only happen within a mental framework where nature (and eventually those 'others living as part of nature' that the westerners called primitives) is the 'ontological other'. That is, the ontological dualism that was preached in the creation myth of the Mediterranean religions (and not necessarily in the origin myths of other traditions, as we learn through anthropological research: Long 1967) established a mental horizon where the world of humans and that of nature are, so to speak, two different realities – where the first one lives on and can make use of the second one in all sorts of ways. As we saw, when 'primitive cultures' were qualified by the first Christian explorers as 'people of nature' and hence part of nature (and not of humanity) they could almost automatically be used as slaves and a dispensable workforce, like animals (Fisher 2017; Graeber and Wengrow 2021). The very fact that the theological position of an often-cited and out-of-the-ordinary missionary like las Casas on the humanness of 'natives' was considered a theme of discussion testifies to this dualism – just as the unending attitude of racism does. But the issue is much larger than humans alone. In this chapter I look at the problem via two categories: ecology as such and freedom and the climate crisis.

Ecology: The Environment or the Biological Context

Until five centuries ago, biology started to study species as separate entities: one became a specialist in the study of plants, animals or of humans. Over time, one specialized even more – to such an extent that my colleague who 'works in trout studies' hardly knows what the department specialist on butterflies in an adjacent office is up to, let alone the one who has gained international recognition for his studies on conifers. Taxonomies defined an order for species and led to the specialization of scientists in one taxon most of the time. In recent decades, biologists have come to recognize that the context of each and every species is extremely important in order to build dependable knowledge. Hence, ecological perspectives – looking at a species in its environment – became respectable. This was and is a sort of awakening, I would say: we gradually became aware that species do exist in relationships to other species and to other, wider and not strictly biological contexts. Today, I claim that in economics and in political disciplines a similar shift of perspective is on the table. The old humanism, with its focus on humans as the only or the exclusively relevant agent, should consider contexts here as well. It should change 'humans as the measure of things' to 'humans-and-other beings as the measure of things'.

Over the past few decades we realized that research had to shift from a species-specific focus in biological sciences, a zooming in on the individual or parts of that entity in psychology and the study of presumably isolated 'states' and 'cultures' as entities in themselves, towards viewing any and all of these phenomena in relationship to their contexts in the social sciences and humanities. To give an example: we have to study ecological relationships between species in the ocean, the sociology of trees and whatnot in order to understand what 'nature' amounts to. In social sciences we have gradually learned that the second territory of Wilhelm

Wundt's nineteenth-century sketch of psychology (focusing on the social and cultural embedding of individuals in the last volume of his magnum opus) somehow 'fell off the wagon' in the following century. The daring approach of the Vygotsky School in the 1920s was not fully picked up in the West until the 1970s, and even then it only survived as a more marginal subdiscipline of individual- and ego-centred research (Cole 1996, on all of this). The comparative perspective in linguistics, sociology and even in anthropology saw a similar fate: exceptional scholars like Franz Boas, Claude Lévi-Strauss or Philippe Descola emphasized the need to compare and to situate the group or community in their contexts (historical, political, cultural, but also ecological), but scholars continue to build their career on the basis of their specialization – i.e. in a particular culture, a state, or an artificially isolated group or community. Still, over the years, systems thinking, the integration of context into the research, and comparisons have proved to be more needed than ever: butterfly collecting is not enough, the one-focus bias is a problem, etc.

The general rule was (and still is to a large extent): cut up the reality you are studying in order to make it manageable. We are increasingly becoming aware that this 'slicing' is not without loss of content or that it even produces irrelevant or fake phenomena. I give an example from my own fieldwork: when I study knowledge and learning processes with the Diné (Navajo Native Americans) I am continuously at a loss about the feature of 'the piece of reality' I envision. The Diné of today are wearing cowboy boots, and up to 90 per cent are trained in western-style schools. Still, a lot of them speak and write the native language and attach great importance to the cosmology found in so-called traditional as well as in modern peyote cults. Within the political and economic context of White ('Anglo') powers, which at the very least co-determine their horizon of survival, they nevertheless proudly resist 'americanization' in the sense of assimilation. But then: what is the 'theoretical or hypothetical identity' of the Diné? How could anthropologists – and, to an even greater extent, educationalists, politicians and missionaries – reason about the Diné as if they were a more or less closed-off, relatively isolated entity unto themselves? And realize that they, along the way, commit the sort of enslavement and other types of discrimination that can be seen in the exhibition in the Washington Museum of the American Indian? Finally, Diné will emphasize that the desert context regularly contradicts the view on the world of the textbooks, used in the reservation schools: resources are not unlimited in a desert, and neglecting or denying the characteristics of the environment will soon harm the people. The story by Dagha Ch'ilii in a former section of this chapter expresses this way of non-schoolish thinking.

As stated before, the world is growing towards structural interdependence today. We witness the fact that choices and decisions in one part of the world impact on the survival of communities at the other end of the earth. Forests on fire in California, Siberia and Australia impact on the quality of the air but also on food supply, ocean life and the traffic in timber in faraway places. In the ecological crisis the world is witnessing today, with the rapid extermination of species and the systematic cutting of tropical forests, it is dawning on us that humanity is rapidly endangering the biological diversity of earth, polluting the oceans and creating so-called 'dead seas' in every continent. The idea that resources are unlimited is shown to be false: tigers, lions and elephants will not recover if we keep this mentality up for a couple of decades more. But also, this continuous attack on biodiversity is backfiring: desertification is a fact when tropical forests are cut, and several recent pandemics are becoming increasingly understood as side-effects of the attacks on biodiversity.

The shift in scientific frames I am mentioning will most probably need to be a deep one. I make this presumption on account of its appearance in quite fundamental domains of research. Some recent biochemistry and micro-biology researchers into evolution, for instance, will at the least allow for a theory on evolution which grants that there is an important margin or rest space for the deterministic version we know from Darwin's theory. Instead of seeing all evolutionary change as the result of 'blind variation and selective retention' only, a group of evolutionists now recognize what is called epigenesis. The latter deals with change processes that reflect types of impact from the context – e.g. learning processes in series of species influence and maybe even direct evolution (see Campbell 1989; Callebaut and Pinxten 1987; Kauffman 2010).

Incorporating the context(s) is precisely what the holistic perspective invites us to do: it lets us see all beings – including humans – as intrinsically embedded in a tapestry of connections with each other. It proves to be unwise or incautious, maybe even dangerous, to look at oneself as detached from all the other phenomena. It will be considered a lack of caution to choose or make decisions that neglect or refuse to take into account the ecological environment. To consider this 'ecological' perspective (I use the term in a very broad sense, to be sure) it is good to use my experience with Navajo–Diné again as a concrete example. Their way of dealing with the world illustrates what is meant; it can easily be compared with several other traditions (as in the works of Descola 2005, 2016 and 2021; Ingold 2016 and 2017; and de la Cadena 2011).

I return to my fieldwork in order to make the point as concretely as possible. In my attempts to understand the relationship of language to cognition, I was dissatisfied with the approaches of logical positivists and

94 • Humanism Revisited

Wittgenstein adepts in analytical philosophy in the period 1960–80. For one thing, these philosophies did not properly differentiate between atemporal and diachronic aspects of language: it seemed that such approaches took for granted that the structural makeup of the individual today is the unique focal point, disregarding how such an individual is formed by deep structural aspects of syntax and semantics over generations. At the same time the methodologies used suggested that some 'type of soul searching' by the individual linguist (inspired by analytical philosophy) could yield dependable insights in the 'universal deep-structural characteristics of language'. Rather obviously, the linguists were 'staying home' in that endeavour and hence based such universal theory on structural analyses of European languages, disregarding the rest of the world (see Hymes 1981 for an elaborate criticism). My naïve intuition, then, was that such structurally very different languages as Chinese or the Athapaskan tongues (considered to be 'verb languages' in a very rough characterization) might not really fit into this frame of reference. Not surprisingly, the question was hardly asked since the linguistics carried out by the descriptive and comparative linguists was not present in the discussion of this new 'scientific' approach (e.g. such rare comparativists as Emile Benveniste [1969] or Dell Hymes [1978] were not heard).

I started out on my fieldwork on 'Navajo thoughts on space' (Pinxten, van Dooren and Harvey 1983). I wanted to know how they thought about the relationships between language, thought and action. I had learned that there is no verb 'to be' in the Athapaskan languages, and that in a sense the then-famous so-called couple of 'noun phrase' and 'verb phrase' of transformational-generative grammar (TGG) did not obtain or at least felt tremendously awkward in the study of such languages. Along the way I came to know Gary Witherspoon (1977), John Farella (1984) and Jim McNeley (1981), who were working on the same sort of problem. They were anthropologists posing deep philosophical problems that philosophers thought to be irrelevant. In this particular case (as on many other questions, see the literature cited above: Descola, de la Cadena and others) these Native Americans lived according to the conviction that speaking and ceremonial or ritual singing is just as impactful as acting. At the same time humans are thought to be so deeply intertwined or interconnected with everything else in the universe that one always triggers and influences the life of the latter, even when silently thinking. Because of this interconnectedness, humans will cause animals and plants, but also the mountains or the winds, to react on the impulses sent out. More often than not, Diné will say, this reaction will be harming or causing discomfort: *nizhoni* (beauty, harmony) will turn into *nixzhoni* (disharmony, illness, death). It is then that the individual who caused this disruption will be sent to a shaman who will diagnose what happened, and will suggest a healing ceremony with a medicine person. The latter will think, speak, sing and act in such a way that the unhappy choice is undone and harmony might be restored (see, e.g., Farella 1984).

So, what do I conclude? Obviously, this Diné way is one particular cultural incorporation of deep ecological thinking and acting. I do not propose that we should all go and 'assimilate' to become Diné. But this and the other 'holistic' cultural perspectives we come across in anthropology point to an intuitive, foundational or pre-rational frame of reference which places humans in a network of interconnectedness with everything else.

In terms of the South American Quechua, they integrate humans in the world of 'living beings' (de la Cadena 2011). Or, to frame it in terms of western dualism, it places humanity and nature in an integrated network of things where interconnectedness is a structuring given. Hence, the responsibility of humans cannot be limited to what they choose or reject in their position as unique or final decision makers (remember the tenet 'humans as the measure of things').

When I defend this point of view as a humanist I will have to rethink and rediscuss some of the tenets of the old humanism (and Enlightenment philosophy): some centuries after the birth of the daring revolt of the old humanists we have to recognize that this European perspective on reality allowed for interesting positive developments (the often-mentioned results of progress) on the one hand. But in the past century it also resulted in life-threatening effects: the destruction of biodiversity, causing pandemics; many 'biblical' forest fires; the desertification of vast territories, with rapidly growing famine; the creation of 'dead seas' and so on. Against these and other effects of short-term, human-centric, reckless behaviour humanists should reconsider: where did it go wrong? Should we build buffers or warning signs into the frame of reference we stick by? Or is it more rational and more responsible to widen the focus and include other species in the club of primary decision makers? In the latter case, should we include other species and hence look again at what we qualified as 'our enemies' in the past – such as insects, the wolf attacking livestock, the birds who 'stole' our crops, 'wild nature' and so on? If we expand the relevant whole to include nature, what place do we grant humans then? This is one discussion that should be started. If we take that road, should we then 'legalize' nature by giving it legal rights? That is what some ecological thinkers have been proposing: grant animals rights, and forests, and maybe the oceans, as well as the polar areas. That would be the western way, I claim that in the wake of Enlightenment thinking (Kant up to Rawls) we tend to formulate human relationships in social-contract terms, which are expressed progressively in legal terms: rights, laws, sanctions, juridical negotiations and so on. The least we should do, then, is to try and convince 'the others', via respectful NE-interactions and negotiations, that this is the right choice or even just a sufficiently sensible and cautious one. Personally, I have doubts: western history shows that privileges and over-juridification have often been the net results of such efforts, while the destruction of the ecological environment has not been stopped. But that is my position in the discussion.

My point here is that this sort of soul searching has to be engaged in. This appeal notably touches upon a related focus in old humanism: the tension between freedom and responsibility. In the recent pandemic these notions have been at the forefront of protests. It suddenly seemed as if the humanistic claim of human freedom was promoted by anti-vaxers, extreme rightist groups and internet trolls, who declared themselves to be under siege by governments and media. Was that indeed the case? At the least these claims should make humanists alert, and hence I devote a section to this point here.

Freedom: What Exactly Are We Talking about?

In a brilliant recent study on freedom, historian Annelien de Dijn (2020) analyses this notion, which played such a central role in humanist and Enlightenment literature as well. Political democracy is based on freedom; economic theory and practice in the so-called free-market tradition take it to be essential; and freedom of choice in religious, social and ethical matters can be found in all constitutions of democratic states as a basic value. The historical reference par excellence is, according to de Dijn's analysis, Athens during the Classic Age. It has become part of numerous history curricula today that Ancient Athens, some 2500 years ago, was the little urban state which was able to break the military siege of the Persians through its political cohesion, based on freedom. The execution of the tyrants brought freedom to the Athenian citizens of the fifth century BC. From then on, they decided via procedures of free discussion by all what political decisions would be made. It is fair to keep in mind that 'all' was meant to be understood as all free Athens-born men – women, slaves and foreigners were excluded from this freedom tradition. Decisions which were reached this way then became the rule for all: that was the essence of democratic decision making. About a century later, important philosophers of the time started criticizing democracy and its idea of freedom: Plato was a notorious anti-democrat, who rejected this notion of freedom. Later, Aristotle was the philosophical tutor of the military emperor Alexander the Great.

Under the Roman Empire, a similar trend can be witnessed: the republic, with its democratic institutions where free speech reigned, was ended by the autarkic Julius Caesar, who founded a long line of emperors. De Dijn shows in her thorough book how diverse Christian figures and organizations basically continued political unfreedom until the humanists of the Renaissance dared to reopen the road towards free choice and free speech. The ideas promoted by them (starting with Petrarca) explicitly referred to the Roman Republic. De Dijn states that political practice showed that over the coming centuries, throughout the Protestant Reformation and Catholic Counter-reformation, rulers would sometimes introduce some of the instruments of democracy in England or the Low Countries. But overall, political domination by the few (nobility and kings, helped by bishops and theologians) over the many would remain the rule. Neither did individual conscience gain a primary position, not even in the writings of Martin Luther or John Calvin. In that same period private ownership (mainly of land) and other forms of capital gradually became the solid foundation of power, alongside or instead of military power or God's word. But freedom did not gain a central position in these times. With the so-called Atlantic Revolutions (i.e. the American Revolution of 1776 and the French Revolution of 1789) the Athenian notions of freedom and democracy re-entered the political discussion. But the steps taken in order to institutionalize these principles were very modest: slaves and women could still not participate in elections, and power remained overwhelmingly the privilege of landowners. De Dijn mentions that even Enlightenment philosophers did not really make a difference here: Jean-Jacques Rousseau stressed the limits of freedom; John Locke was satisfied when state power was limited, democratically or not. With the Bill of Rights of 1791, the author detects a shift away from the Athenian model that would last for centuries: individual rights gained a higher status at the expense of shared democratic decisions.

Some remarks may be added to underscore de Dijn's deep study: in his continuous study of inequality over the centuries, political economist Piketty (2019) convincingly shows that private ownership (of land and/or capital) can be seen as the main discriminating principle in the arena of political decision making, laid down in constitutions and in the bylaws of all nation-states, notwithstanding the Revolutions.

The historical analysis by de Dijn shows clearly how the notion of freedom has been picked up and used in ethical and political instances through the centuries in a variety of ways. The linking of freedom to democracy has not been obvious, and even less commonly intrinsic. Most of the time, since the Enlightenment, freedom has come to stand for 'liberal freedom' or, in the old distinction, for 'negative freedom': I am free when the chains of Church and king are broken. Put differently: I am free from obligations. When translated in terms of rights, this then comes out as: I am free to choose X or Y, since I have the right to choose unhindered by others. The Athenian view of freedom in democracy, however, was different from this merely negative programme: I can discuss and choose freely in continuous interaction with my fellow human beings (citizens or family or whatever), but when we reach an agreement by majority or consensus, then I side in solidarity with the group's decision.

In the recent publications of Graeber the relevant anthropological literature adds to this analysis. In his study of democracy Graeber (2013) relates how the Founding Fathers of the USA, at the time of the Revolution, did not even aim to start a democracy. On the contrary, most of them were suspicious about this political system (propagated in the French 'salons' of that era). Instead, they wanted a well-organized federal model that would be efficient. Furthermore, the latest book by Graeber and Wengrow (2021) shows how elaborate discussions on freedom of choice may have been the reason Native Americans were very reticent to be charmed by the much less free society of the White colonial powers.

If we adopt a (more) positive notion of freedom in order to have a democratic society we have to decide which responsibilities will be equally binding for all participants, even if their adoption implies that the ideal of 'negative freedom' (freedom from all ties) will have to be seriously amended. In this book, that is what I confront the humanist tradition with: looking back at the effects of our reckless and impactful way of dealing with others (the NE-aspect) and with nature (the E-aspect testified to by the loss of biodiversity and the climate warming), I propose discussing what responsible choices for freedom can be made. The mere reference to private, unlimited choices of individual humans will not do: the premise that humans are the measure of things apparently has not prevented the endangerment of our species and of much of nature, nor has it guaranteed the freedom of all. On both points, we can certainly highlight some historical progress for a minority of humans (the 'haves') but, all in all, the ideal did not work for everyone – let alone for nature. Humanists should, then, reconsider the responsibility issue and seriously discuss how and to what extent individual freedom can be reformulated in terms of responsibility for fellow human beings and for nature. It is not so much the community level that matters here, but the crucial limitations that would have to be agreed upon at the global level. If humanists do not take that road, they are in fact acting as defenders of a new version of exclusive and irresponsible tyranny. The label 'Anthropocene' is often mentioned in this regard: the era when humanity has become the owner-director of all on earth.

Beyond the historical study of freedom, however, there is a second and equally urgent concern that we should mention here: the way the shift towards an IT society is taking place. In itself, the use of IT in all sorts of technology and facilities should not bother us. But the way it is 'taking over control' should be on the agenda of any humanist. I treat this point here, since it may be appreciated that the impact on the value and the practices of freedom will soon become alarming. The COVID-19 pandemic is showing, to all who are willing to see, how the IT economy is changing basic political and ethical choices – mostly beyond the conscious, let alone critical, choice of its users. So-called social media has proved to be very instrumental in spreading false messages, in establishing 'alternative truth' and in popularizing conspiracy theories. When on top of that extreme rightist parties in the West (Ponsaers 2021; Höhne and Meireis 2020) on the one hand, and a set of powerful conservative governments on the other (with President Trump in the USA, but also colleagues such as Boris Johnson in the UK, Jair Bolsonaro in Brazil and so on), allow, approve or join the unwarranted use of this means of influencing, the very notion of freedom is under attack. It is to the credit of Shoshana Zuboff that she has devoted a thorough analysis to this issue (2019). Her book's title carries the message: 'Surveillance Capitalism'. The author shows how the IT industry recently, somewhat to everybody's astonishment, worked out what enormous profits could be made by what is now called 'data mining'. First Google and Apple, and later on other players, understood that the internet allowed immense amounts of users to communicate at very low cost, leaving stacks of data on their life choices, their ideals and dreams at the mercy of the providers. With a simple 'agreement' we all learned to allow 'cookies', and thus transferred the right to use these data to the providers.

They started looking for likenesses in preferences and dreams, and made groups or communities of them, which are then sold for great amounts of money to industrial partners. The latter thus got hold of unknown and physically intractable 'communities' who share tastes and opinions and thus constitute ideal target groups for the products of their businesses. In a second moment such 'grouping' and targeting was then tried out for political use: Cambridge Analytica was probably hired by the US Republican Party during the presidential campaign of Donald Trump, with a view to influencing sets of undecided voters 24/7.

Pretty soon, influencers and salespeople learned how to act in this virtual space: 'alternative facts' appeared as a category, types of stalking of uncertain or scared people boomed. In a further step criminals and political extremists detected the dark net, which allowed them for several years to escape from any democratic control. There the 'bubbles' of target groups could be manipulated even more. In her very systematic and hair-raising study Zuboff gives a tremendous overview of the way the IT bosses gained control over people's minds in no time. Not only did they build up an economic empire that way, but they started weighing on democracy as well: for example, when the CEO of Facebook was repeatedly called to explain his firm's violations of privacy policies (by the European Parliament and by the US Congress) it proved that his way of working in the market was not so much violating as bypassing existing democratic rules. Moreover, freedom was narrowed down in the minds of the users almost exclusively to the right to say and do what one chooses, beyond the control or the limitations formulated by any state or international organization. Paraphrasing old humanists, the tenet then read: 'freedom is *my* being the measure of things'. On top of that, through practical developments in the field, the very notion of freedom seems to evaporate with the deployment of IT facilities, pushing society towards what Zuboff calls 'surveillance capitalism': we are henceforth traceable 24/7 through cameras in the street, but even more so through mobile phones and other devices. Moreover, for a few international firms, the ways to manipulate people in their choices and preferences are almost unlimited: in practice, bottom-up agreements on working conditions and participatory decision making in industrial concerns are on the verge of breaking down, yielding a new type of 'slavery' (extremely low payment, no social security, etc.). At the same time, media groups and advertisement concerns have an ever-more powerful influence on markets and on politicians (funding their campaigns, for example) such that the notion of free choice is gradually emptied out. In China, meanwhile, a direct and straightforward authoritarian version of Big Brother control is installed through face recognition and suchlike, allowing the government to reward or punish citizens at will: social security benefits, the possibility to study, the right to live and move around in a particular region and so on are linked to the continuous IT control of each individual.

In that system, of course, the notion of freedom is abolished in a straightforward way.

Conclusion

This ends my analysis of the second dimension of an updated humanism, as I see it. My invitation here is similar to the one I sent out in the previous chapter: inclusion is the basic value I want to promote. Inclusion of human diversity should be enlarged by the inclusion of all 'Earth Beings' in the scope of humanism. We are only responsible choosers when we allow and actively guarantee that the interests of other species as well as those of other important natural phenomena should be at the basis of humanist thought and decision processes. The complex system of humans+nature should be the measure of things, not just human beings alone. We have learned that we need this inclusive emphasis lest we ruin the existence of other aspects of the world and, finally, our own chances of survival.

In a final step of this reasoning (D humanism) I will expand the horizon even more.

Chapter 6

D

Durable Humanism

Durability, sustainability or resilience is the topic of the final letter in this new anagram (NEED). Durability or sustainability are used here primarily in the political–economic sense. Of course, the Renaissance thinkers were not aware of the scientific findings and concepts which came in the ensuing centuries. The very fact that most of the humanists and their successors in Enlightenment times were either members of the Church or part of /in the service of the nobility probably made them relatively far removed from common people. I conjecture that these individuals, however brilliant they were in their time, would be rather bewildered by the world of later centuries. Their emphasis on critique and freedom of research remains valuable, though, provided we reformulate and update the several concepts.

My appeal is, as stated before, to integrate awareness of environmental, political and cultural parameters as relevant variables: time (as change), other species, climate, social and economic factors. As a consequence, notions like freedom, individuality and the unilateral focus on 'man' (as the measure of things) and suchlike would have been analysed and phrased differently in these earlier times. In the slipstream of which sustainability may become an intrinsic part of humanism-cum-Enlightenment. At the present moment in western history, such changes of perspective imply that models and theories have to be altered accordingly, leading to 'paradigm shifts' or so-called 'revolutions' in science (Kuhn 1962). Since these shifts do not follow from the tenets and practices of the old humanists – for whom the free will of the individual reigns supreme – present-day humanists are quite reluctant to adopt them.

That the perspective is too narrow in the humanistic and Enlightenment defenders of today can easily be argued when looking at the present heroes in the field, such as Pinker or Dawkins in the Anglo-Saxon world: theirs are typical 'histories of ideas', and the dimensions I work out here are hard to find in them. I will only mention some of their work occasionally.

What is durability or sustainability? In his lesson on sustainability the anthropologist Tim Ingold (2017) elaborates on this subject. A first instance, covering some but far from enough ground according to Ingold, is that of management culture: it is characterized as metric durability. That is to say, one is thought to be busy with sustainability when one counts how many entities of a species or specimen is available. On the basis of such counting one then decides how much to keep in stock (e.g. keep resources in the ground), to reduce or to multiply in the next period. Think of a supply firm selling tables: depending on what is still available in the stock, it can be decided that new acquisitions of tables need to be postponed or augmented. This sort of reasoning, says Ingold, is appealing to many because it works in numbers: the modernist loves to reduce the complexity of reality to data. This is not a simplistic statement: think of many debates about forestry today, even against ecological arguments. One will count how much CO_2 a forest is able to process and then proceed to offer the reintroduction of a certain amount of seedlings, which will of course produce new wood to cut in as short a period as possible. At first sight, the balance seems to be taken care of: we cut all trees and replace them with the ' 'correct' number of trees that will digest a similar amount of CO_2. The ontological dualism referred to in a foregoing chapter thus proposes transforming quality (a forest, a whole world of diverse plants and animals) into quantity, i.e. the amount of CO_2 digestion. What is gained and what is lost? A similar example is the actual international politics of 'emission quotas': Belgium has way too much air pollution and hence 'buys' clean air by giving subsidies to Mongolia, which is hardly industrialized and hence has on average 'too much clean air'. Belgium hands over the money on the condition that Mongolia will not industrialize, and it continues to pollute the air in Belgium for another generation. That way, Belgium sticks to the international agreement of 'reducing air pollution' – at least, in the bookkeeping. Quantitatively speaking, the balance could be restored on a worldwide scale; qualitatively, children in Belgium are stuck with the ill-health consequences of high air pollution, Mongolia cannot reach a higher level of wealth and the over-pollution in and through consumerism in the West is not addressed at all. The trends of the destruction of biodiversity and the changes in climate are manifestly not taken seriously, as worldwide statistics show (van Yperseele, Libaert and Lamote 2018).

Already in Ancient Roman times, a natural philosopher like Lucretius (see 2008) had pointed in the right direction here: nature (or maybe broader 'reality') deals with how phenomena persist and continue to exist in the cycles of life. For example, when one takes such a life stance seriously, it is quite obvious that the cutting down of an age-old oak tree which is living in a 'society' of other trees, bushes and whatnot, and then planting fast-growing poplars in order to reach a similar amount of CO_2 capture from the air, is an imposture. A great deal of quality of life for all is sacrificed for the benefit of quick gains for some.

Going back to the issue of forestry, Ingold (2017) relates the case of a Japanese community cutting wood in order to build houses. The procedure is as follows: each tree is actively raised and protected for about thirty years, after which it is cut down and used as building material for a hut. Meanwhile, a new tree is planted and 'nursed' to replace the former one. After another cycle of thirty years, the hut is destroyed and replaced with the wood cut from the tree of the latest generation, and so on. The house is in a sense the 'second life' of the tree. The cycle is the relevant focus. Raising and cutting are two moments in a continuing cycle which expresses sustainability as the primary quality. Also, trees and humans are partners in the cycle, which cannot be understood as just an 'external' type of management in the hands of humans alone. A similar, though concretely different, attitude I found during my fieldwork with Native American Diné: in the delicate and obviously vulnerable balance of the desert context in which they live, simple harvesting according to the economic logic described above is considered dangerous and thoughtless. In contrast to White people, Diné would never extract all specimens of a species of plants or exterminate all members of an animal species. Their overwhelming emphasis on the restoration of balance between everything (humans and every other phenomenon in nature) testifies to this concern. When their Tribal Council decided to develop a plantation of vegetables and corn of all sorts along the San Juan river, this view on things was dominant. The calculation was that enough food would be produced to guarantee the survival of most of the Native Americans at that stage. In market terms this sort of approach could not compete with the agrobusiness farms of the USA, especially the large one-culture plantations of Arizona and the Midwest. It would suffice, however, for the survival of Native Americans, independent from that market agriculture (which is subsidized anyway). However, stacks of law suits followed, amongst other things accusing the Diné of 'crypto-communism'. The report on this experiment, which was dubbed 'red capitalism' by the circles of Republican presidential candidate and then Governor of Arizona, Barry Goldwater, can be read in Gilbreath 1973. This initiative of the Diné, which would today be qualified as a tribal 'common', was crushed by powerful political and economic groups at the time.

A final example from anthropological research focuses on a very different example, turning on a recent development in refugee policies in Europe. The French anthropologist Didier Fassin has been working with groups that have become victims of discrimination: Black people in South Africa and boat refugees in Europe – especially in Calais, France, whose refugee camp has been dubbed 'the Jungle'. In line with the philosophy of Michel Foucault he investigates the existential and political opinions about 'life' that circulate in such groups. In practice Fassin (2021) distinguishes between an ethical and a political interpretation that come to the forefront: life is sensible, either meaningful or empty, depending on the group under scrutiny. Fassin observed that a very expensive surgical intervention on a Palestinian child sponsored by an Israeli benefactor became world news, at the 'right' moment for western political opinion makers. During the same time sequence a group of many children and adults were knowingly left to die in a rubber boat on the Mediterranean Sea, where they have been followed by various ships (amongst them NATO observation craft) for several weeks. The 'incident' was hardly mentioned in the news. Ethically and politically these are different notions of 'life', Fassin claims. I mention this long-term study of his because it demands from every scientist in such fields that she looks beyond the mere numbers and takes into account the broader contexts which co-determine what is a relevant or integral part of the question under study. In the examples cited the humanist–researcher will undoubtedly support the life-saving activity for the child. The careful researcher, however, cannot stop there and ignore the international political context that is forcing its way into the equation. Even if the benefactor did not know about the role played by other ships, including the military forces on the Mediterranean Sea, the researcher cannot possibly be satisfied with the problem formulation when it leaves out the deadly effects of the policy referred to by Fassin.

This sort of example, about natural phenomena (such as forests) and about people, is not exceptional or exotic in our time: we tend to know, or at least are able to know more than ever before, what the contexts are which co-determine fate, life chances and welfare. When that is indeed the case, it is of the greatest importance for the humanist to integrate the broadest and strongest notions of durability or sustainability when developing the life-stance values to be cherished. Being satisfied with individual values and choices, or taking the human being as the measure of things, will unfortunately not be sufficient.

In order to appreciate what this amounts to I will now go into some deeply entrenched contextual dimensions that were not, or were insufficiently, present in the consciousness (and in the conscience) of the old humanists or the first Enlightenment thinkers, and are still absent in the humanist proposal today. The economic context as well as the dualism of

106 • Humanism Revisited

humanity–nature, and its consequences for thinking and choosing, are two such dimensions.

'It's the Economy, Stupid'

This was, of course, the famous expression attributed to the President of the USA, Bill Clinton. Looking back I can say that the president underlined a premise of neoliberal ideology by picking this particular emphasis.

In his recent book *Internationalism or extinction*, the famous linguist/political thinker Noam Chomsky claims that sticking to this ideology will lead humanity into the abyss in no uncertain way.

What should we understand here? One needs to be clear about what is called economy to begin with. In a very general sense economy is nothing more or less than the set of principles, values, insights and practices which allow a household to persist. It is the set of all means used to detect and fulfil the needs of a group or community (a family, a region and eventually a state, and so on). And economic choices focus on the way needs, means and procedures can be integrated to secure survival. When I cite Clinton and Chomsky, the 'household' rules and practices I refer to are those of states and even of international and global entities (companies, government agencies, markets). My point for this book is that questions of choices, values and their implications when taking economic measures have been considered too scarcely in the humanist tradition.

Of course, human beings live as individuals, meaning that they have self-consciousness and thus can make choices and decisions that engage them as unique persons. But, as stated before, human beings are formed by and add to two other levels of existence: they are members of several groups with interpersonal contacts, and of one or more communities through virtual contacts. The latter extends today without any doubt to the global scale. In the humanistic perspective, the focus was almost exclusively on the individual agent as the seat of value decisions. This was quite understandable as a reaction against ages of domination and uniform enculturation in Europe under the aegis of Christianity, where individual decisions were reserved for the top strata of the community. However, as has become clear in the present time through the crises mentioned, blindness towards group and community parameters leads to egotism and the endangerment of nature and humanity in our times. Or, in other words, it yields what is called illiberalism: this proves remarkably true with an ideology which goes under the misleading name of 'neoliberalism'.

Still, even this criticism is too shallow. I follow Amitav Ghosh's (2022) historical anthropological analysis, where he traces back the shift towards 'economism' and the ideology of neoliberalism to the early colonial period. In the Portuguese; Dutch; English; and, finally, the American way of thinking about the world in terms of 'extractivism', whatever was found in the world from the voyages of the seafarers onwards was seen in terms of profit. In contradistinction to many native views on the coasts of the Indian Ocean and in the Americas, nature was seen as inert, dead and consisting basically of resources to be extracted and used to enhance the wealth and dominance of the 'clever' westerner. The latter reduced nature (including the 'natural peoples' or 'primitives' who often had an enchanted view of it) to rich resources that were there to be mined and used at will by the colonizers. Ghosh has it that fossil energy became a fundamental driving force from industrialization to our own days because the dominant groups in western societies understood that this fossil energy (coal and oil, primarily) presented a unique combination: on the one hand, its extraction allowed for gigantic profits; on the other hand, it promised the never-ending dominance of those who were private owners of the resources over those who were employed to mine and process them. The development of military systems throughout the world by European colonial powers, and later by the USA, went hand in hand with the extraction history, Ghosh (2022) shows. This analysis critically assesses capitalist developments, the author claims, but the fact that anti-capitalist powers (like the USSR and China) continued along the same track makes the trend more general by now. That is why Ghosh prefers to speak about a general worldview of 'extractivism' in its different political–economic guises.

As is clear from earlier chapters in this book, I see an accompanying root in the 'ontology' that is typical of the Mediterranean religions of the book, later inspiring economism: the dualistic ontology with nature as the inert entity to be used without limit by the only 'willing', ethical and hence rightful user/conqueror, i.e. humans. Any 'holistic' ontology was thus progressively pushed back and almost eliminated as 'primitive', irrelevant in the presumably universal progress that the West was thought to instantiate.

As shown earlier, some anthropologists are increasingly developing a critical reappraisal of the self-glorification of this approach to reality in general, and of the predominance of economistic foci in particular. Moreover, it is in critical economic theories that at least a minority of economists and business leaders can find the most pointed criticism. That is why I deal with them first.

In meticulously documented long-term studies Piketty (2014) proves that since neoliberalism came to power in the 1980s on a global scale, inequality has been increasing sharply. In a more general way, some old Christian and new humanist values seem to have landed in the 'dustbin of history':

108 • Humanism Revisited

- Solidarity and redistribution is systematically fought against: competition and individual initiative are highly praised and sustained in education, the sciences and the arts, at the expense of solidarity and shared interests.
- 'Greed is good' is promoted as a major value, thus allowing for tax evasion in a massive way in the past generation. New icons of 'success businesses' of the IT and new industries (IKEA, Tesla and so forth) pride themselves on paying as few taxes as possible.
- Privatization of almost anything is guaranteed through international negotiations: forests are privatized, the air can effectively be bought (with emission agreements), water and land and the resources of the soil are claimed by corporations (e.g. the battle for the North Pole resources) and even human organs and DNA can be 'bought' or patented by clever agents of Big Pharma (Mattei and Nader 2014). The voluntarism of the individual humanist shows no power whatsoever to counter these developments.
- The prevalent existential value sounds more like 'live NOW, since YOU have the freedom to do so'. The emphasis on 'now' leads away from thinking in durable terms. The focus on individualism only promotes selfishness and lack of any broader responsibility to encompassing values, thus adapting (or is it corrupting?) the humanist critique (see the anti-vax movements of 2021–22).

With the 'heureuse trentième' solidarity and redistribution became important values in the decades following the Second World War (1950–80), but the neoliberal period which came afterwards destroyed many of the structural and institutional results of that period and allowed the rise of a new elite, which became ever-more selfish. Early on, this shift was accompanied by the promise of the 'trickle-down effect': when the elite became richer, their wealth would somehow automatically trickle down to the benefit of the lower classes. The reality we have witnessed since the 1980s proves different: wealth is concentrated in the hands of the few, up to the point where they will sponsor so-called research centres to produce 'alternative facts' which detract attention from the real state of the world (Stiglitz 2011). The fact that wealth is increasingly extracted from the earth's resources and from labour by the lower classes is documented in the brilliant analyses of Thomas Piketty (2014 and 2019). When the president of one of the world powers (President G.W. Bush) weakened many international treaties by having translated the bulk of international relations in terms of trading agreements (between real or so-called business partners), most poor countries were caught in a trap of inequality (Mattei and Nader 2014). UN agreements no longer protected them from usurpation by multinationals, since they were engaged in trading contracts now and not in rules of international law under the UN umbrella. It was another illustration of selfish and short-term political choices, yielding benefits for the wealthy and hardship for 'the others' inside rich countries and elsewhere in the world.

The ideology that corresponds with this deep change can be understood as promoting less-democratic concerns and even anti-democratic action (cf. the many contracts with dictators, to guarantee cheap petrol, gas and other natural resources). Added to this emphasis one finds a turn towards economism instead of economy. Economism is a belief system – i.e. the belief system that free, unhampered market mechanisms are economically and even ethically superior to all other projects. Of course, a few critical economists pointed out that this switch will land us all in a blind alley (Stiglitz 2019). Or they would emphasize that economic theory is deeply entrenched in values and interests of the agents involved: 'I dislike the expression "economic science", which strikes me as terribly arrogant because it suggests that economics has attained a higher scientific status than the other social sciences. I much prefer the expression "political economy"...' (Piketty 2014: 573–74). In this way critical economists position the ethical and political choices and the value-ladenness of economic theory against the ideology of economism, where value choices and premises have gained the status of unquestionable dogmas – very much like in the Mediterranean religions. These critics are a small minority, though. In that 'religious' format of economic theory the range of values and possible choices appears as 'given beforehand', as an axiom – that is, regardless of broader contexts and in line with the benefits of the elite.

This rather long introduction should allow discussion of sustainability in economic matters. Linked with this point I will expand in a final section on the 'social-contract' thinking emanating from the humanism-Enlightenment tradition.

Staying within the premises of economic theory of the past few centuries in the West, one can stick to the rule of market thinking in a narrow sense: the dominant balance at any moment is that between need and supply, and contract rules regulate the details. In most domains of economic activity, short-term profits influence the decisions taken to a large extent. A mantra that is often heard today is: technology will solve problems wherever and whenever they show up. Long-term thinking, and certainly the costs of entrepreneurial initiatives to that effect, is notably absent in an overwhelming amount of activity: dumping waste massively in the oceans and/or in poor countries has become a rule, and leaving the costs of 'side-effects' to nature and/or those who do not act as economic agents is widespread (air pollution and other health issues are left to the community, climate change will be at the cost of all, etc.). In that respect the latest report of the

NIC (2021), which I discussed earlier, breaks away significantly from the usual reports. This representative body for the American intelligence centres presents a synthesis of the long-term effects of past incautious policies and invites everybody on government services and in private economic organizations to sit back and look at the prognoses that are on the table: the NIC presents a set of scenarios for the future. The easiest in the short run allows for doing as little as possible, and repairing some damage (when storms hit and so on). The most engaging scenario proposes that more responsibility will be necessary on the part of economic agents, reducing climate problems drastically and opting for ecologically neutral production and distribution as much as possible. The latter implies less so-called free market, since the costs of choices by market players should be calculated in the long term and should be honestly integrated into the business balance: no dumping free of charge any more, no pollution, no pillage of natural resources.

When one sees that this change of perspective is being suggested by the official scientific and governmental organizations of a major free-market country (the USA), it becomes clear that the general mentality might be changing at last. I mean to say, it is becoming absolutely clear that long-term thinking from a perspective of global responsibility is increasingly seen as unavoidable, or at the least the more cautious and hence safe way to go. This implies, as the NIC report and other sources (van Yperseele, Libaert and Lamote 2018; Hickel 2021; etc.) show, that the scope and maybe even the content of the notion of (individual) freedom is being problematized. The minimalistic and individualistic notion of humanism, as it has been legally framed in the social-contract ideas and structures since the Enlightenment, safeguarded a rather negative freedom: I have the right to choose and act independently, since I am free of obligations and rules installed by the powers that be (Church, king, etc.). Certainly in the Anglo-Saxon version, this was understood as the road to individual free choice, with the smallest role possible for state or community regulations. The way I read reports like the one by the NIC, alongside studies by the IPCC and other scientific boards (and recently, even the OECD and IMF), leads me to think that a necessary shift is at least setting in – away from the purely individualistic (egoistic) and short-term notion of 'freedom from' towards a situated and positive set of values of freedom. The latter stance reads: at the limit, my freedom implies responsibility for long-term effects and for the world beyond me. This is a difficult shift for many, and could mean a deep change for major economic agents. In my view, it is high time that humanists–Enlightenment supporters join the discussion by reconsidering and deeply adapting their old views. The knowledge that humanists first and foremost revolted against a corrupt dominant hierarchical power system is not only a given of the past but might also inspire the humanist of today to stay alert about community or state power, but it is becoming clear that depending exclusively on individual freedom is yielding irresponsibility. Confronted with the effects of the incautious and pillaging history of the past centuries under the aegis of 'free-market' views, together with the 'negative freedom' premises, it is high time to develop and integrate notions of responsible freedom in an interconnected world of humans and other natural phenomena. This redefinition could remain loyal to the ideal of breaking away from hierarchical obedience, which was so central at the time to the revolt against Church-cum-nobility domination. It should, however, creatively work on freedom-with-responsibility, and I suggest that here we can learn from the horizontal, interconnectedness view we find in other, non-western traditions mentioned before. I review a few issues we will find on our way when choosing this approach in order to revitalize and update humanism–Enlightenment thinking in the changing world we are experiencing today.

As a side-effect I expect that only an appealing, updated life-stance programme along these lines will be able to turn around the growing anti-humanist, irrational and indeed conspiracy-minded movements we witness today. Still, my conviction is that the latter may not be a passing accident, since major political parties use or protect them already – e.g. the Republican Party of the USA, possibly deeply under the influence of rather extreme groups of fundamentalists; movements like 'sovereign citizens' in all western countries; and suchlike. Similarly, several leaders and political formations in other powerful countries around the world, from China and Brazil to Russia and Poland/Hungary, support the anti-humanist trend. From the side of so-called liberal or social-democratic countries and parties, uncertainty and endless (and often unclear) compromises abound – aiming, most of the time, at short-term solutions for the problems of the day only. Climate and ecological problems are addressed in international conferences (e.g. the COPs), but the resolutions reached there are rather systematically not followed up when the delegates return home. Also, the growing inequality of the past few decades is not halted in any efficient or systematic way by the rulers that be, although the risks to the survival of all are strikingly clear by now. Of course, rethinking humanism-cum-Enlightenment traditions will not solve any concrete problem today, but it may offer a frame of reference and a mindset on which to base future decisions. For one thing, indifference and anxiety may diminish as a result.

Interconnectedness and Universalism?

It should be clear that universalism is not to be understood as 'universally given'. In its present form universalism may be part of a deep problem, since it proves to be highly culture-specific. It is part of the colonial or Eurocentric mentality that we need to get rid of (Hari 2023). How can we understand this?

As is well known the 'universals' that philosophers have been talking about fit within a frame of reference of the Christian tradition (and probably more generally that of the religions of the book) and have a very particular status. In the literature the few attentive critical thinkers always mention the full expression: 'universals of transcendental philosophy'. As is rightly argued by Amartya Sen (2009) in his famous and highly ambitious alternative to this long western tradition – including Kant and Rawls as Enlightenment thinkers – universals have an a priori status, which is believed to make them true in the sense of unavoidable or superior regardless of empirical arguments. In that sense they are 'given': an alternative would deny reality, so to speak. The best-known example, since it has been referred to extensively by nearly all legal and political thinkers for over two centuries, is that of Immanuel Kant: when he started thinking about the building blocks of a necessary and sufficient theory of knowledge, he 'logically' and on the basis of consistent reasoning came to the conclusion that three basic, a priori, universal categories had to be identified. They were: the concept of the existence of God, and the notions of space (in a Euclidean understanding) and time (in the spatialized form of time as an arrow). This sort of reasoning was loyal to the Christian a priori *universalia*. God, space and time (in the particular format they have in this part of the world, and with the linguistic structures that prevail here) are, as it were, emanating in reality from the way God decreed them in the word he gave to humans (called 'Revelation'). Their universality need not be checked, let alone proven, and can be relied on to be valid for all because it is obvious or unavoidable. Consequently, when Chinese mathematics (and other versions of the discipline in other cultures) proved to be very different, one was not worried: these 'others' had it wrong.

Looking a little more into the Chinese tradition, it is striking that it developed an elaborate algebra but no geometry (Needham 1965–2000). To my mind this at the very least may be compatible with the fact that classical Chinese has a predominance of verb forms, with almost an absence of the noun category. Put rather bluntly, in their thinking the world is one of events, processes and not of things. For the lay person: the famous yin–yang couple is not a combination of two 'things' but the ever-rotating and alternating interactive process of two dynamic principles. At the same time it is impossible to deny that Chinese technology, architecture, hydrology and alchemy (to name just these few branches of knowledge) were so well developed that the Chinese Wall could be constructed by means of them, or that irrigation works over hundreds of miles were successfully helping millions in agricultural development ages before the Europeans started thinking about the possibilities of such technologies (Needham 1965–2000). Hence, western explorers of China in the sixteenth to eighteenth centuries imported or borrowed a great deal of this knowledge (including that on fireworks, yielding gunpower in the West). Similar 'unorthodox' knowledge and 'strange' technology have been detected more recently by archaeologists in several non-western contexts of Antiquity. For example, Graeber and Wengrow (2021) discuss such examples from diverse parts of the world, dating back to 10,000 and more years ago. Apart from the fact that these 'primitive' peoples left unmistakable proof that they were able to use reasoning and intelligence, the way several of these constructions were erected puzzles researchers to this day. These peoples did not seem to be bothered by what Europeans declared to be the particular yet exclusive *universalia*, but nonetheless produced efficient knowledge and technology. On top of this, recent studies in ethnomathematics offer a variety of insights which diversify and enlarge what was held for a long time to be the only true way to knowledge by westerners (Vandendriessche and Pinxten 2023).

One way of opening the debate is to question the a priori or transcendental status of the concepts and categories held to be the only serious, and hence 'universal', ones. This is attempted by some researchers – especially of a Latin American vintage – today, who speak about 'pluriverse' and 'pluriversal' (e.g. Escobar 2017). As a political act, this may work. But it is obviously rather superfluous if one does not go into the tedious work of somehow demonstrating/proving that other frames of reference are equally as valid, powerful and hence warranted as the western one. At a deeper level problematizing the universalism claims of the western traditions opens up broad and very interesting avenues to think differently, but also to interact along different roads with one another. I will concentrate on this topic here.

The problem we are facing, then, is that universalism may be used as the correct concept but, simultaneously, humanity seems to share traces, habits, and ways of thinking and acting that are species specific rather than culture specific and can be recognized at a global or worldwide scale. At yet another level, one clearly finds so-called cultural diversity between different communities, recognizable as thoughts and habits that are typical, local or part of an identity ideology and that are used to distinguish the said community from neighbours and foreigners. Also, human beings within a community or tradition will stress relative likenesses or differences of humans vis-à-vis other species. At the limit the 'otherness' of other species can vary from a difference in kind, and thus an absolute distance, to one of nuance. A cultural community can feel connected in a variety of ways, from almost no to almost complete relatedness. Western religion-based dualism can be situated in the first case: there are two realities, so to speak, that are radically different though living together in one world. There is the human reality (with mind – hence reasoning and language – hence morality – hence rights: Descola 2005), and there is inert 'nature' (which is said to lack mind – hence language and reasoning – hence moral dimension – hence rights). The opposite position is that of the horizontal holists: they conceive of one reality, wherein each creature or phenomenon is deeply interconnected with and impacting on everything else. Western universalist philosophy can clearly be situated in the first version while it is an oddity in the second perspective, and vice versa. Not surprisingly, humanists and Enlightenment thinkers were entrenched in the first option and did not manage to look or think beyond this option. In a sense, they were caught in a 'prison of the mind': embedded in their religious beliefs and incapable of dealing with divergencies in an inclusive way, they kept neglecting – not hearing or even misrepresenting – the 'others'. They saw themselves as belonging to the unique, superior civilization transiting into the world of the saved (a godly, ordered step beyond the heathen state of humanity), later dubbed that of the civilized or developed part of humanity. The very few exceptions who escaped more decisively from this parochial frame of mind – such as Michel de Montaigne – were later marginalized or almost forgotten. In my understanding, the so-called system builders (René Descartes, Immanuel Kant, etc.) disqualified them as relativists, 'bad thinkers' and so on, which is strikingly similar to the insult used against the sophists by that much-acclaimed consistency prophet and rightist thinker of Antiquity, Plato (Toulmin 1990; Perelman and Olbrechts-Tyteca 1957).

So, when I stage 'interconnectedness' on a worldwide scale as an alternative to universalism, what should it look like? And how can we start preparing people to at least explore this diverging road? The second question is the easier one: anybody who ever worked with people belonging to any type of category of 'the others' knows that a primary handicap, as a researcher and as a human being in general, is that western education did not prepare us to think diversely. Even more so, we were not trained to interact with people in ways that differ from the standards of western middle-class persons: we are not used to listening to people who talk differently, or apparently hold diverging values and preferences to ours.

Again, a concrete example: when trying to explore how Navajo people conceive of the spatial environment, I first tried to work as systematically as possible. That is to say, I made lists of Diné words that had to do with space, orientation, movement, cardinal directions and so on. I then went over them one by one, which is a tedious task, inventing as many sentences as possible wherein one or the other term could be used. Structural means (e.g. taxonomies, synonyms, etc.) were suggested where I ventured they might be appropriate. Several informants worked with me, sticking to the task at hand as closely as they deemed necessary, and my Navajo interpreter kept track of the whole endeavour in close collaboration with me. Then he urged me to go and speak to one particular person, who was working at the time for a social-service organization on the reservation. I went up to him and spoke about the general outline of my research: within the deep structural differences of this Athapaskan language I was looking for the way spatial relationships and space in general (remember: one of the three a priori, and hence universal, categories of the major western thinker Kant) were expressed. At first the man refused to work with me, and sent me away on some excuse. Days before I would leave the reservation after almost a full year of research, he sent for me. He then started to explain what he thought this should be all about. He started telling a story, using some of the terms I had been investigating but clearly refusing to use the so-called descriptive models (taxonomy and suchlike) that linguists and anthropologists at the time were so fond of. In a sense he presented the topic in a story, then went on with another story that seemed vaguely related but could eventually be understood as altogether different, moved on to a third one and finally came back to his initial track. It was as if I had asked for the way from Albuquerque to Los Angeles, and the man had started out to go to Chicago, then to Rome, then to Moscow and suddenly arrived in Los Angeles. All the while he kept looking at me, as if to check whether I was still with him. He presented two such journeys as answers to my single question: 'let's talk about space, the Navajo way'. Of course, he was aware of my research up till then, since he knew my interpreter well and had worked with anthropologists intensively before. He summed up: we can answer your questions the way you (Anglos, White people) are expecting it, 'but that is not the way we Navajo think'. The Navajo, oral way of sharing knowledge is an interactive, person-to-person way of exploring questions. To this day, I am grateful to him for trusting me that way. Indeed, I had already moved away from the structural descriptive methodology myself: it proved efficient and easy to work with, since as a researcher one could keep a constant overview of data and structural relationships between terms, but the frame of reference of the informant's culture seemed to escape me – and so the data felt skinny, or at best trivial in a sense, and their interpretation was basically left exclusively to the researcher. The more I experienced the way of life of my informants, the more this 'data collecting' became unsatisfactory to me. Concretely, I had experienced how young children (of age 5 or 6) were sent to roam around in canyons with a sheep herd and a dog, for two or three days. The children had only pre-school knowledge. They got home safely in a vast territory without roads, signs or any other 'Euclidian' or data-like means. I would be totally lost, and indeed would risk my life, by starting on that kind of journey. They 'experienced' the landscape, the sun's trajectory, the rare watering places for the animals, the places to spend the night safely and so on. They worked with a mostly implicit but englobing cosmology, as expressed in the orientation of the rare *hooghans* (houses) they came across, the appearances of particular plants in certain places, the behaviour of animals, the many 'movements' and changes that are continuously captured in the 'verb language' that is Navajo and so on. This knowledge was clearly adequate in the untamed territory they wandered around in. In school, two or more years later, they were then urged to drop all this 'illiterate knowledge' and replace it with western, literate curriculum material (including Euclidean notions) lest they fail at school. The dependable knowledge they gathered by listening to stories and through their own experiences was declared useless, primitive and therefore obsolete.

I decided in later fieldwork to take the pre-school knowledge seriously and use it in the educational process in the classroom. Working with Navajo teachers we developed a 'Navajo mathematical/geometric vocabulary and frame of notions', which they 'constructed' in a couple of secluded gatherings with fellow teachers (Pinxten et al. 1987; Pinxten and François 2011). This curriculum book would then be used in the maths class: starting from pre-school knowledge more formal thinking ('mathematics') is built up.

This example shows what today is recognized in the various more-systematic approaches of 'multiverse' thinking (e.g. Escobar 2017) or different 'worldings' of problems and insights in a variety of ways (my translation of *mondialisations* from Descola 2021), rather than the one, exclusive way of the western, transcendental philosopher. This point of view need not imply that 'anything goes', however. In the words of some South American scholars, elaborating the basic *convivir* perspective, different emphases and diverging models can be construed depending on the local history, the 'natural' environment (e.g. different for a desert, a forest or a mixed vegetation context) and the way people interact with their particular surroundings.

Recognizing the plurality of perspectives on things – especially as illustrated in the different cultural outlooks we have become aware of since travellers and, later on, anthropologists made reports on many cultural traditions – remains a difficult issue for us, humanists–westerners.

In the wake of exploration, the economic conquest of riches from all over the world was swiftly framed in terms of rights: the right to own what we found on and in the earth. This focus on 'right', and especially on private-property rights, gave birth to an enormous rationalizing tradition:

philosophy to a large extent forgot about Montaigne's honest amazement and human interest when confronted with subjects from unknown territories. Instead, theologians and philosophers fell back on the Eurocentric idea of the universalism of ethical and knowledge claims from just one tradition, and promoted the European human to be at the forefront of civilization for all of humanity. The first critical remarks by the newly discovered peoples themselves were soon forgotten or pushed aside, when the thinkers of the capitalist–colonialist powers took over. Lately, some anthropologists have begun criticizing this imperialistic approach towards 'others'. Eric Wolf, for example (1981), has pointed out in detail how 'Europe' gradually framed all other cultures as ahistorical, lacking the 'right' view on time and hence on 'progress'. Johannes Fabian (1984) made a philosophical critique along the same lines. Laura Nader (2015) published a bundle of historical texts (dating from the seventh to the twenty-first century), written by members of other traditions who had learned to deal with Europeans during these centuries in diplomatic encounters or in business deals. Recent studies are feeding this eagerness for alternative, diverging, indeed pluriversal perspectives (Graeber and Wengrow 2021 and Hari 2023 – but also earlier in Fabian 1979). In terms of durability in the political–economic field, these types of studies may help westerners to at least start thinking differently about our ways of treating others and their perspectives on human survival.

The notion and ideal of inclusion allows for pluriversalism instead of 'universalism'. It follows remarks on what are also called 'other worldings' (i.e. ways of viewing and expressing reality: Descola 2021): the latter will no longer a priori be qualified as wrong, underdeveloped and therefore not worthy of serious attention.

From there on, a series of possible conclusions follows:

- The transcendental approach of western philosophers is seen as a local one: these thinkers started from what they thought to be 'evident' or necessarily valid on the basis of their local intuitions or premises. The ensuing, impressive philosophical constructions they formulated, mostly on the basis of internal coherence and consistency, may be interesting, ingenious and whatnot, but they now prove unsatisfactory to be labelled 'true' on the basis of the a priori notions advanced. For example, Kant's a priori category of space as necessarily being what Euclid based it on, since it is 'in essence' the only possibly right view for humanity, was already under attack a generation after the great thinker recorded it, when a trio of playful mathematicians (Bernhard Riemann, János Bolyai and Nikolai Lobatchesky) developed so-called 'non-Euclidian geometries'. The Non-Euclidiean geometries started from counter-intuitive premises and proved on the basis of these what was impossible or 'false' in Kant's view. Two attractive a prioris were: 'the shortest distance between two points is a curve' and 'two parallel lines cross at some point'. Their geometries now prove more convenient (more true?) when working in ballistics, or in astronomy, than the awkward or 'local' Euclidean version.

- The intuitive sources of inspiration which lay at the basis of thinking and searching, and from which the premises for reasoning emerge, can differ substantially. Deep, structural differences in language and thought, and diverging contextual possibilities and constraints, yield often clearly distinct worldviews. In the European/western view on reality, humankind and nature are situated on either side of a principal divide: reason–choice–ethics–legal rights on the human side versus mere mechanical processes–instinct–amorality–no rights for 'nature'. The structure of this reality is hierarchical, with humanity in the 'driving seat' of the one who controls, uses and decides. From the intuitive perspective of *convivir* and similar traditions, humans are caught in an intrinsic and pervasive network of relationships with all other phenomena. Interconnectedness, and hence the sharing of survival values, is intrinsic to this 'horizontal' and holistic view on reality.

My claim is that only now, and sparingly, does a critical scrutiny of deeply felt premises stand a chance. It is in that respect that I look at the growth in many circles of such rather uncommon inter- or transdisciplinary research as taking short-chain economics seriously (see, e.g., Raworth 2017); rethinking the dogma of growth or progress (Hickel 2021); looking at a forest as a grouping of social, intercommunicating communities rather than a set of 'trees-as-CO_2-disposers'; looking at the wolf in the total context of an ecosystem rather than as a competitor for meat with humans; and so on. The list is growing daily. My invitation is that NEED humanists should venture actively into this self-critical turn, and base deeply modified life-stance values and rules for conduct on it.

The discussion on such premises is still in an initial phase, it appears, since scientists remain overwhelmingly trained in the usual competitive, short-term and even monodisciplinary frames of thinking and choosing – e.g. psychologists or sociologists hardly get any substantial training in theories and models other than the western ones, nor do philosophers on the whole know anything substantial about knowledge and ethical traditions other than those of Ancient Greece and what came afterwards in the West. Students in economics focus overwhelmingly on marketing and applied economics, and know next to nothing about other survival traditions. The curricula are not deeply changed because the mindset is not either. Of course, changes are called for today, but it may take at the very least another generation for these changes to become more mainstream. Nevertheless, deep changes in perspective (including on the ontological level) are setting in, and humanists should join this trend and reset their lore.

Does all this imply that 'they' know as much and as adequately about reality as 'we' do (with our science and technology)? The question is as yet unanswerable. How would and could we compare? Which concepts and which arguments can we use for such questions? But 'we' realize from experience that our reckless approach, based on the dualism referred to, drives us toward the 'sixth extinction' – so, a degree of modesty is in order.

Preparing –– albeit hesitantly – the next generation for a more diverse and possibly pluriversal view of survival strategies is the task that some philosophers and anthropologists define for themselves now: this is the explicit intention of scholars like Graeber and Wengrow (2021), Descola (2016) and some others, as it is for the author of the present book. In the light of the present crises such an engagement appears to them as decent, if not necessary.

In everyday practice, politicians, thinkers, economic agents and religious leaders keep steering the world. Their inspiration and their justification will keep coming from the ideologies, belief systems and other sets of values and norms that are offered by contemporary humanists. Hence, we should start to develop alternative views to the transcendental philosophy that has been so typical of the West, and which has remained solidly entrenched in the humanity–nature and us–them oppositions. Only then may humanism become a genuine perspective for humanity's future, in close relationship with natural phenomena.

Continuous Social Choices: Listening, Negotiating Instead of Governance

It is in concrete policies that power is enforced most of the time, or occasionally negotiated. Against the background of the critique I have worked out so far, it is now time to look at possible alternative views and procedures. Since the reigning imperialistic approach to 'globalization' calls for more profit for some and more disaster for many, I call upon humanists and Enlightenment adepts to look for this deep change of perspective. With the knowledge that humanity is factually more and more linked in interdependence relationships it is time to leave behind the supremacist attitude and look for other ways to interact with all of humanity, as with all other phenomena on this earth. Admittedly, this is a very ambitious call. But my feeling is it is a call out of necessity. The change has to set in, in order to survive the crises which humans have called upon themselves and the rest of our planet during the past generations of ever-more global extraction and pollution.

Once more, I start with a concrete example. Two decades ago, I was asked – since the study of non-western cultures was the common definition of my discipline, anthropology – by the Flemish chapter of UNESCO to formulate a comment on the draft of the upcoming Convention of Heritage (and, more specifically, immaterial cultural heritage). The meeting was followed by a small reception, in the presence of the young Queen of Belgium. So, I gathered there was some sort of official status attached to it. The text we were given to comment on had been 'amended' through lobby groups: this was not the story we gathered from the officials, but it transpired after a short debate on some of the clauses. Let me be systematic here. The text had been discussed in several national councils already. It said that local customs; thoughts; and, most of all, copyrights had been safeguarded, in line with the UN tradition of respect for all cultures and nationalities. In some countries this was read as a legal means to protect the local cultural production against powerful international media groups. In particular, it was admitted that France wanted a clause to allow for French film and television producers to protect their own 'markets' against the Anglo-Saxon big players. It was admitted by the chair of the meeting that Anglo-Saxon groups, under the leadership of the Disney concern had then lobbied for free access to any market worldwide for their products, notwithstanding rights of production for each and every country. In other words, the markets for distribution had to be safeguarded by lifting them out of the semantic field of 'protection of immaterial heritage'. This, then, was the deal between continental European and other players in the now-profitable business of 'immaterial heritage'. Some countries or regions (including my own) wanted to accept that deal provided their local/national heritage production would be granted identity status, fencing it off against intercultural influences. This implied, as I remarked as a novice in these sorts of meetings, that each country or region would have to define a solid and island-like cultural identity, which would be protected against any foreign influences. This should be done through the identification of one's particular immaterial cultural identity, regardless of humanity as a whole.

Looking at it more closely as an anthropologist, this proposal generalized the mistaken framing of 'others' as cultural islands which developed and closed off their identity as different from those of neighbouring communities and, in the end, excluded something like a shared humanity. In that view, of course, humanity as a common given or a shared necessity is not recognized as a significant feature of reference for any separate community. At the same time, it was clear that the powers of the time would benefit most from the compromise: they would defend their rights against less-powerful, i.e. southern, players. That is exactly what the colonial mentality had promoted as a premise in order to establish its rule over 'the others' for as long as it lasted. Short-term reasoning, not durable. Moreover, as I tried to make clear to the council, this is precisely what extreme-right groups and parties, supremacist thinkers and identity movements were claiming in their continuous us–them mantra. Some colleagues granted that they, as 'scientific voices' on the matter, had over-looked this aspect of the choice they were about to approve. Some of my philosophical and sociologist colleagues at the gathering were angry with me: what could an anthropologist – 'a specialist in primitive cultures' – tell us about an agreement between modern states? They pleaded that we had to reach an agreement with universal validity, amongst nation-states My point was: maybe, but how could that be reached if one stayed stuck in discriminating choices to begin with, based on a short local history, even if one was not really conscious of that?

In the period that lies between then and now, we had 9/11, the wars in Iraq and elsewhere, the shifts towards a world divided between maybe ten powers if not more (as announced by Kissinger 1994), now engaged in a new war in Europe with no end in sight. On the other hand, in my understanding at least, it becomes ever-more undeniable that the world is increasingly interconnected and humans are now factually interdependent for survival. Hence, reverting to cultural-identity solutions is a form of 'insularization', which is so in conflict with the factual interdependence of the twenty-first-century world that it cannot possibly claim durability.

Chapter 7

WHAT NEXT?

Good Guys and Villains?

When giving talks about the questions raised so far, I often run into a wall of misunderstanding – sometimes even reproach. Christian humanists (a minority from secular continental Europe, a majority from the USA and the UK) will say that they are doing good whenever and wherever possible. For example, with the clash between Ukraine and Russia starting in the first months of 2022, they very rapidly and efficiently organizing structural help for refugees from that area: governments and Christian Democrat parties in the European Union accepted refugees with almost no thresholds – offering beds and even long-term sponsoring for them to come over and stay, as well as a place in the economic system. They were granted permits to stay for at least two years. This is 'doing good' for any humanist, my audience will tell me. Yes, of course, but does it not strike anybody that refugees from other origins (Afghanistan, Iraq, Black Africa) are simultaneously being kept under horrible conditions in detention camps at the European borders (Greece and Turkey, for example) or in even more doubtful camps in Libya, subsidized by the EU? If anything, this is selective charity: we are humane with the white Christian refugees but much less so with Muslim refugees of colour, a critic will conclude (think of the work of Fassin 2021). On top of that there are obvious pushbacks.

Another example: the secular or 'freethinker' humanist in Europe will typically object to my analysis with the following type of argument: what is wrong with defending equal rights for all? Should we stop fighting for the good treatment of homosexuals in countries where they are persecuted and discriminated against because the local religion condemns their choice in life? Or should we look away from the atrocities of certain tribes against their enemies, and from the discrimination of women in their midst? What do I, the critical examiner of humanism, have to say about the faults and crimes of the 'others', and why would I object against continuing to stress the importance of free research and free speech?

Here again, I run into objections and reactions of annoyance and irritation that are of a defensive nature. This is understandable, but nevertheless rather shallow. My analysis is not meant to annoy, let alone offend anybody. And it is certainly not meant to deny that a lot of people have good intentions. But voluntarism and good intentions are, unfortunately, not enough. Even worse, they may hinder or harm structurally 'good' changes. That is exactly my uneasiness with the bestseller by social researcher–journalist Rutger Bregman (2020): *Humankind: A Hopeful History* (*De meeste mensen deugen* [*Most people are decent*] in its original Dutch title). Bregman claims, 'Most people are decent' (in Dutch this is even more voluntaristic, since he refers to the 'good intentions' of the majority). I do not share the belief that good intentions or charity will suffice, even if they are shared by a majority. Indeed, along with Arendt and others, I want to point to the many instances where actual discrimination is happening under our noses, and is passively or actively condoned by this same 'majority of good people'. The genuine humanist, the one who stands for universal freedom and equality and freedom for all of humanity, cannot look away here and continue to stress that 'they' (e.g. the Muslims) are discriminating themselves (e.g. against women or LGTBQ+ people) or that uttering this statement suffices in itself. It is not enough to condemn a vice in somebody else's behaviour and then look away from a very similar one in one's own. Also, it is far from noble to keep on defending morally 'good' principles when one's actions are demonstrably not aligned with them. History has shown (and here I side again with Hannah Arendt) that such an attitude produces new discriminations and crimes, showing lack of courage and principle rather than high ethical standards.

So, reference to what 'others' do or did wrong (and thus making 'them' the villains) does not make 'us' any better humans and heroes, let alone our human traditions/communities. And the other way around, pointing to misdeeds, hypocrisy or outright wrong choices in our past does not make the 'other' holy and pure either.

Does that mean that all is relative, and any attempt to look for a decent and dependable basis is false, necessarily ending in a nightmare? Of course, I do not have an answer to such a question. I can say, though, that humans have shown a sort of capability that has not been discovered so far in other species: we are fallible but at least have been repeatedly conscious of our fallibility. To format this feature, we have developed the weak but real 'culture of critique'. That, at least, is a remarkable piece of human creativity. This formidable but vulnerable cultural product (in the sense that it is not 'natural', but it can be grown as an offspring of human creativity) can be spotted in periods of peace and welfare in different parts of the world: for example, the Tao tradition in China seems to qualify, as well as some North and South American native historical traditions (Graeber and Wengrow 2021). It is certainly a basic principle of humanism in the West and part of the standard scientific ethos. Maybe it can be spotted in all of humanity's divergent cultural traditions, once we start looking for it, because apparently it will take a diversity of forms and shapes. My point, then, is that many ethical positions can be reasoned on in a similar way: we need a decent comparative study before we draw conclusions of a so-called universal scope. One forceful, concrete rule for the future we might then decide on, and try to spread through global negotiations, is to always reserve room for fallibility.

In focusing on the Western cultural tradition, therefore, I can obviously detect the one remarkable value that has been explicitly debated and – during short periods of western history – safeguarded as maybe the best guarantee against crimes, harmful mistakes and structural injustice against others and/or nature: that of the self same 'culture of critique'. This is not so much that of the quasi-theological system thinkers (the so-called 'great' philosophers) who fell for the textuality trap and mainly developed intra-textual critiques. My plea is to always safeguard room for the doubters and the relentless seekers. When reading Chaim Perelman and Sylvie Olbrechts-Tyteca (1957) or Stephen Toulmin (1990) it becomes clear that for short periods in our history we managed to allow for all possible criticism of the ways of a community, a religion or a political doctrine. Toulmin calls this the second stream of reasoning in western philosophical history: it pops up and stays around for a generation or two, and is subsequently shuffled under the ground again for a longer time. The sophist thinkers can be identified as an early example (sixth–fifth centuries BC), opposed by the anti-democrat Plato (and Aristotle) and later by the institutional thinkers of the Christian era. The humanists in the fifteenth and sixteenth centuries could be identified as another upsurge of this root of radical, critical thinking. They were neutralized by the religious warriors only decades later, yielding to more or less orthodox system builders. Maybe some radical Enlightenment thinkers – those inspiring the so-called Atlantic Revolutions (de Dijn 2020) – were another case in point. Their trend was also silenced, this time by upcoming capitalism and its empire building and by the nationalist ideologies that came along with the latter. Finally, some thinkers of the post-Second World War generations could qualify too (Thomas Kuhn and Paul Feyerabend for theory of knowledge, the Frankfurt School for ethical and political thinking). But every time, say Perelman and Olbrechts-Tyteca and Toulmin

What Next? Good Guys and Villains? • 125

in unison, this critical tendency was overruled by the powers that be: today, the neoliberal philosophy of extreme competition between individuals (a new version of Hobbes' philosophy?) applauding the value that 'greed is good' narrows down any critique on ethical and political fairness, durability and suchlike to questions of efficiency in a managerial way. The mantra of 'knowing is measuring' is back in power, and management reigns over open-minded questioning.

Whatever else, our attention to this vulnerable 'second stream' of thinking (Toulmin) will make us conscious of the fact that it is not something that is acquired or established once and for all. At the same time, these fallibilists and self-critical thinkers were not 'negative' or relativist thinkers who just enjoyed tearing great systems down. Rather, they were mostly 'possibilists', to use a term which was promoted by the late Chancellor of the Free University of Brussels Caroline Pauwels (2021). When allowing criticism from a perspective of fallibilism, you can withdraw into cynicism or you can go and try to reach a better solution or a more humane agreement by negotiating with all involved because you see an opportunity, a possibility to reach a better society, better knowledge or more happiness. That is the principle of possibilism, linked to fallibilism, as explained by Pauwels. It joins the appeal for global, mutually respectful and less 'rationalistic' thinking about rights and values of Amartya Sen (2009). I gladly join this line of thinking: it may result in more modest, less grandiose systematic philosophy (away from the Kants, Augustines and Rawlses of western tradition), but it will try to engage all (also the 'others') and everything (also 'nature') in an honest and fair negotiation to come to satisfactory choices for survival.

If we follow Toulmin's analysis of the attitude of respect for the 'culture of critique' it proves rather exceptional in our history, and maybe in some other histories, wherein obedience to authority (heteronomously to a God-figure, or autonomously to some human ruler or doctrinal institution) is much more common. What I recognize in some humanism–Enlightenment endeavours at least is the onset of such a 'culture of critique' as it blossomed in the historical context of five centuries ago. However, the hesitant steps of these forefathers in the European context of that time somehow came to a standstill: the critique of heteronomy stopped halfway and the self-critical analysis of western superiority – on the pillage of nature or on almost holy economic principles such as progress, mechanistic thinking or private property – was only very rarely continued, but never became mainstream. The good news may be that negotiation with others, and a different perspective on nature, may finally stand a chance in the present era. That will only become true, I think, when we take fallibility seriously in the wake of self-critical humanism, albeit in a context where we threaten to drive humanity

126 • Humanism Revisited

in a final abyss if we keep looking away. As possibilists I believe we can make the right choice here, and hence should start working on programmes such as a modest NEED humanism. As a realist I have to recognize that the deep crises we have landed in and still refuse to look in the face do not constitute the ideal motivating force to start such a change in perspective.

My Ideal

An updated, positive and engaging humanist–Enlightenment project will have to reconcile the said factual global interdependence for survival, the emancipation from Eurocentrism, the recognition of rights of other phenomena (than human beings) and the primacy of durability in economic and political terms as guiding principles. That, at least, is a minimal programme for now.

Many questions come to mind when stating such an ideal. Just to give a taste:

- Is this not a new, only slightly disguised, formulation of a universalistic programme? No, but why not?
- How are we going to reach such a shared agreement? Don't we need to have a sort of 'grand scheme, worked out in the cockpit of spaceship earth'? No, and why not?
- In the same rubric: we westerners know through our scientific knowledge that, and how, things went wrong: the 'others' don't. Wrong.
- Is not the western value of progress (and growth) a necessary condition for any kind of change, even towards a degrowth option? Probably not.
- How can humanity accomplish this deep shift? That is, what sort of education and what sort of ethical choices are imminent in order for humanity to have a future?

Of course, I am not going to formulate a code of conduct for humanity. I will focus primarily on my own tribe, for one thing: the westerners. What I can and will try to do, though, is to offer some insights that may help all of us who are concerned about present developments to combine insights and honest hope in order to start building durable alternatives.

Chapter 8

RESET OR THE EXTINCTION OF *HOMO SAPIENS*?

Wednesday, 6 January 2021: a scene from a second-rate science-fiction film? A barely structured flood of people, a so-called mob, storms the Capitol in Washington, DC, USA. The Capitol may be considered one of the most prominent symbols of western democracy, in a country that has figured as an example worldwide for a type of representative democracy in a recent timespan of two and a half centuries. A lot of those present posted themselves as 'freedom fighters' and brandished banners and flags to underline that identity claim. Cries to incite the 'lynching' of sitting representatives of the people in Congress were spread across social media. Some of the journalists who were present had to run for their lives, and a few people were killed in the uproar. Police forces fled and tried to avoid major damage. The President of the USA did not call storm troops to stop the raid on the parliamentary building. Weapons were all around. After demolishing some of the interior, breaking through the police defence force and occupying the seats in the Houses for a short while, the mob retired from the building. The President of the USA, Donald Trump, who had failed to be re-elected to the highest office, first encouraged the mob without ever recalling them and then withdrew in the background where he started acting as the 'comeback kid' for the next elections. Throughout 2022–23 he keeps on announcing his political re-entrance, encouraging the right wing of the GOP to support this 'campaign'. In the sensational hearings about this period it proves to be the case that the attack was more a 'coup' than a spontaneous uproar of individuals or groups. If this holds, then we have to face a steep problem: in one of the richest countries of the world, the unofficial

128 • Humanism Revisited

leader of the so-called free world (the USA) witnessed the bypassing of any form of responsible, humanity-wide let alone earth-wide concerns or values. This would be more than alarming. The hesitant reaction of most media, which are usually prone to broadcast any possible 'scandal' even on shallow foundations, is not reassuring either: such political derailment (as seems to be the case) only happens in times of deep and possibly structural crisis in the world's economic and political institutions. Hence, a deep and reasoned analysis should be central to the news and to the accompanying public debate. If anything, all media of any ideological vintage should be engaged in this process and be evaluating such deep questions as the way forward for democratic values, the market-driven exploitation of people and of natural resources, and the supremacy of the economic agenda. Indeed, it is precisely these aspects of what can be considered the primary means towards a decent and respectable life in the West that have led to humanly deeply unfair, ecologically irresponsible and even earth-damaging ways: the primacy of 'greed is good' for the elite may be the deep source of the derailment that is visible now on the political scene as well. Moreover, for me at least, the event in Washington, DC (and the ensuing conclusions emerging on the reactionary side of the political spectrum, within and beyond the Trump administration proper) is not a local phenomenon: elections in the present era throughout the democratic world have seen the rise of rightist and extreme-right parties and movements, including in such bastions of democracy as France, Italy and the Netherlands. While looking for an alternative humanistic programme, I will try to focus on some of this in this final chapter.

The attack on the Capitol happened at a high moment of the pandemic COVID-19. Amitav Ghosh (2022) rightly mentions the fact that the most pronounced western defenders of neoliberalism, i.e. the USA and the UK, have been shown to be among the poorest performers in handling the pandemic that has plagued the world since 2019. Proportionally, the highest number of casualties was registered there, due to the poor preparedness for the pandemic and inconsistent messaging. Moreover, the causalities were highest in the poor, ethnic-minority segments of society (especially Native Americans: Ibid.). The privatization of insurances and healthcare systems has made these segments land in unusual situations (at least for the West): the sight of mass burials in New York City have citizens from elsewhere thinking of war scenarios, steep casualty numbers on Native American reservations remind at least some observers of structural segregation. Moreover, these numbers show at a time when drought has been striking many of the same areas more and more over the past few years. In most other western countries virologists became the new stars in the newsrooms of radio and television. They tried to guide the population at large in order to adapt their behaviour to the demands of the situation: lockdowns, social-distance rules, restrictions on public life and so on. Governments in most countries gave the virologists a lead role by inviting them to occupy centre stage: time and again, ministers declared in the media that they followed the lines set forth by the scientists. The top medical advisor on appropriate behaviour vis-à-vis COVID in the US was sometimes attacked by the political elite. Since the enforcement of restrictions on freedom of movement and of social and cultural life, together with a hidden increase in poverty through a serious setback in economic activity, could be seen as essential features of political decisions to be made, the use of scientists to underscore them was sometimes almost dominant. The scientists repeatedly stated that scientific knowledge is in essence fallible, but this clearly did not always meet with the required understanding in particular circles. The 'alternative truth' theme, repeatedly promoted by populists in power, found willing ears in large groups of anxious and suspicious people, meeting each other on the new social media. Old extreme rightists and new religious-conspiracy prophets seemed to find each other in no time through these media, playing on the fear and distrust of many (Ponsaers 2021; Höhne and Meireis 2021). It was in this context that mass demonstrations of people refusing to take a vaccine (the so-called anti-vaxers) and small rioting groups tried to unsettle governments, along the way sending death threats to virologists through the internet. I read distrust in the humanistic perspective here: rock-bottom beliefs rather than fallibility reign.

Following this long and difficult period of pandemic, we then landed in a war in Europe: the first war with superpowers since the Second World War. Russian troops surrounded cities and started bombarding and indeed occupying Ukraine in February 2022. The reactions were 'interesting': extreme-right 'friends of Putin' (the Putin regime had sponsored extreme-right parties in all European countries over the years) and extreme-left parties who still refer to Russia as the heir of the Soviet Revolution of 1917 disqualified themselves by their loyalty to President Vladimir Putin and his government. The EU saw an unsuspected internal cohesion emerging in the wake of its solidarity with the Ukrainian population. The period is interesting, I dare claim, since it seems to have occasioned a profound growth in consciousness on survival issues where years of scientific reporting and public protesting had worked along these lines with only partial success. I mean, the climate crisis made clear to all who could understand that the dependency on fossil resources had to stop. But denial, indifference, egotism and maybe even laziness came to the rescue of the multinational corporations, which did and do supply the material. With the Ukraine war it suddenly becomes clear what this dependency on fossil energy means:

the sky-rocketing prices of coal, oil and gas, imported into Europe in large amounts from Russia, showed beyond any doubt that the dependency on such resources should stop as soon as possible, since we were now experiencing what it meant in practice to be at the mercy of the Russian supplier. So, where science and activism did not succeed in reaching the majority, fear of being in the cold and the dark convinced governments to step up efforts rapidly. Of course, it remains to be seen to what extent this effect will be convincing when the conflict is eventually resolved. But at least for the time being the war woke up large groups in the West, it appears. The hesitant and very cautious attitude of the USA in this conflict can be attributed to military risks, but it looks as if self-interest (with the USA as a supplier of gas and of weaponry) is an important argument as well.

I think it is sensible to draw on these developments in order to invite well-meaning citizens from many layers of western society to take a basic but updated humanist perspective seriously. My conviction today is that the basic message of humanism–Enlightenment is manifestly less widely supported than before, and the groups mentioned (mobs, extreme-rightist voters and conspiracy groups) fail progressively to identify with that legacy. Instead of blaming it on these groups, an update of the legacy is called for.

Universalism Reconsidered

As I explained in former paragraphs, the universalism notion we have known and used for centuries is a normative one. Thinkers applied the rules of analysis to the best of their knowledge, reached what they considered to be lowest-common-denominator conclusions and then 'universalized' the latter to stand as the only ethical, reasonable or otherwise acceptable way for humanity in general. Previously, I mentioned how the a priori categories of notable humanist–Enlightenment thinker Immanuel Kant proved untenable later on: his notions of God, but also of time and space, were discussed in depth for centuries. An interpretation of this history, which I side with, entailed that the theological a priori status of such categories came under attack. Rejecting their a priori status resulted in doing away with the universalistic claims. Put differently, one cannot say they are universal – and claiming that status nevertheless is a political act. One then forces local or particular values and views upon others.

It follows that universalism can alternatively be held when conceived as a 'universalism a posteriori'. That is to say, one that will hold those truths and/or values that follow from painstaking, comparative, empirical research and whatever presuppositions or foundational perspectives are shared by a vast sample (at the limit, all) of people in this diverse cultural reality that we call humanity. In the search for such universal knowledge we will have to be critical about our own deep-seated intuitions. It is good to remember that the second part of the study of humanity in Wundt's general programme was not really developed so far – the comparative study which would make psychology genuinely scientific (Cole 1996). In none of the other humanities or social sciences was the Eurocentrism of the nineteenth century overcome. Hence, our learned views on humanity remained steeped in local premises which we have declared to be of universal significance. Anthropology, meanwhile, was understood to be a discipline of 'other cultures' only, not humanity in general.

It has become clear, in the wake of recent anthropological studies on concepts of nature and humanity, that holistic thinking might best be conceived as fundamentally inclusive thinking and acting: it is not (White) 'humans [that] are the measuring rod of things' but rather different cultural versions of humanity – or, even better, 'the interconnected totality of nature and humans, where each and every partner in this whole impacts on all others' is the best candidate to serve as the measuring rod. If human beings from around the planet could agree on the relevance of this intuition for thinking and acting, then the factual interdependence that we are living today could gradually be translated into universal attitudes by means of which a respectful, 'holistic, buen vivir' (Lotens 2019), or living in unison with all of reality, will become possible.

To make this point clear in a rather straightforward way, it is good to look at some of the recent philosophical discussions in several disciplines (strongly in anthropology, gaining power in other humanities and social sciences). Ghosh (2022) makes an analysis of what is really untenable in the western, and now globalized, approach to reality. He insists that the western dualistic ontology (humans next to/against nature) allowed for what he calls a generalized attitude of 'extractivism': inert nature did not have a will, hence no morality and hence no rights (as Descola [2005] concluded). Therefore, it was not granted subject status and became a mere object in the western ideology of extractivism: it could be used, changed, redesigned and 'mined' in any way imaginable by entrepreneurial humans. Historically, it proved to be first and foremost westerners who conceived of nature as territory to be conquered and used unidirectionally for one's own profit. Westerners developed an extensive system of rights and legal regulations for some competing human agents to underscore this privilege. Thus, peoples around the world were conquered and 'civilized' – or often simply killed off – because they clearly did not share this perspective on humanity and nature. But many thousands of other species were also eliminated, to be replaced by the ecologically harmful domestic animals and the uniform cash crops of the West, both raised under artificial conditions. The same happened with plants which were not recognized as 'useful' (i.e. promising no or not enough profit), rivers, seas and even total landscapes. The industrialization of the past three centuries offered more powerful means to expand this view and change the earth in every possible way.

In the first centuries of colonization of the world, this way of thinking led to the use of many culturally different human beings as part of inert nature: they became private property (slaves, workforces) of the enterprising westerners for a few centuries (see Fisher 2017). After the official abolition of slavery, their situation did not ameliorate in a deep way: it is at the very least remarkable that the great majority of former colonized populations are still suffering the most hardship today (Ghosh 2022; Piketty 2014). Where discrimination is thus still intrinsic in the conceptualization and the treatment of other humans, extractivism most certainly is the uncontested attitude vis-à-vis non-human nature.

It now looks as if the dependable knowledge that we developed about nature over the past few centuries never deeply questioned this perspectival stand of extractivism. All in all, the validity (or maybe even just the relevance) of that stand was taken for granted. Along the way, then, instead of building knowledge by comparing western and 'other cultural' intuitions and the ensuing cognitive contents in a broad negotiation endeavour, what these 'others' could come up with was qualified dismissively as 'primitive'. In other words, it was seen as a somewhat retarded (pre-Christian) or less-developed form of the western, which was believed to be objective and universal in nature. Some minor findings or technologies of 'others' were adopted (or sometimes stolen), like information on what is edible or not, insights into fauna and flora of the area one was about to conquer or specific technological products (the catamaran, gunpowder, etc.). Part of the drive to accept extractivism and the moral values that go with it lies in the belief in continuous progress: the one who is free to take what is in inert nature and process it into material goods and services is believed to work for the growing wellbeing of the exploiting society. Progress itself is an undoubted value, as if a given, in the market thinking, available to those who play the market cleverly. In order to have the latter work for more progress, this market should be 'free' – i.e. subjected to extraction for those who own the resources. In a sense, these beliefs agree with Christian liberal political economy (Hari 2023; Sen 2009). It was only when, on the one hand, (inert) nature started 'striking back' in the form of climate changes, pandemics and suchlike, and humans, on the other hand, began to be less manageable because of too much inequality that the presupposed universal validity or superiority of this view on earth and human species started slowly to come under scrutiny. In the wake of the emerging criticism the intuitive positions of some of 'the others' have gradually been getting more attention today. Humanism of the twenty-first century can and should take the lead here, in my view; in a sense, it will simply pick up the critical process it started five centuries ago but failed to fulfil when capitalism became dominant. One necessary step in this second dawn will be to develop genuinely comparative ways of seeing and thinking.

Universalism a Posteriori:
How Are We Going to Reach Shared Humanistic Stands?

When discussing such issues with younger people in interpersonal contacts or through internet forums, I increasingly encountered the remark that the lack of appeal of humanism in contemporary western countries is a fact, just like the decline of current democratic institutions and political parties. The opinion seems to be spreading that the slow and cumbersome path of debating, negotiating and looking for compromises on important issues in a world in crisis produces fundamental doubts. When the house is on fire, it looks inappropriate for many to start discussions and reach compromises: quick and firm centralized action would be more effective. So, younger (and also some older) interlocutors objected to my invitation to reset humanism: 'what we need now, if only for a short while', they would claim, 'is a sort of enlightened dictatorship'. Autarky (heteronomy) will save the planet, not humanism's autonomous decision making. Some of the discussants would even point to the possibility of another 'great dying' (Ghosh 2022): let a strong leader organize the world in such a way that the people in the most threatened and actually poorest parts of the world die out now. Keep the borders of the rich parts of the world (the West primarily) tightly closed to keep these needy peoples out in an efficient way. Both governments in the USA and the EU should act 'more firmly' against refugees and 'illegal' immigrants than they did a few decades ago. Most probably, some interlocutors tell me, that will have to be done by 'strong men', not by democratic rulers. After a couple of generations, the argument goes, the world population will have been reduced sufficiently that the threats to the climate will be 'manageable' again. In fact, some so-called 'realists' will tell me, that is what we are doing already: we keep out the poor climate refugees, who flee their countries today because of changes in climate and hence in a survival perspective. So, grow up, stop dreaming and join the rebuttal movement.

The poor regions of the world are manifestly paying the highest price today (Ibid.). So, why not do it in an open, efficient way? And able dictatorship will stand a better chance of achieving that goal than stumbling, democratic compromises. I grant that these anti-humanistic opinions may sound extreme, but the fact that they are discussed seriously points to the present disbelief in the legacy of humanism.

Of course, when these arguments are voiced the humanist needs to address them. This can be done in a series of ways, I think:

- *Outrage*: the call for a pause in humanist–democratic rule in order for a tyrant to get things back on track is not new. When looking at the Ancient Greeks one recalls that the choice for democracy was ended with Pericles, justified by the anti-democratic rhetoric of Plato and his ridiculing of the sophists. The same happened later in Rome, when Julius Caesar seized power. And much later in western history, with Nazi Germany in the twentieth century. I grant that in none of these cases was the outrage of humanistic inspiration sufficient or even remotely adequate to counter the course of events. So, outrage is unlikely to solve the problem.

- *Dictators have always been more of a problem than a solution*: I do not know of any example (including the short periods of so-called enlightened autarkists in eighteenth-century Europe) of dictatorship that did not end sooner or later in corruption and utter disorder. Many great plays by Shakespeare and other playwrights, stacks of political poems of all vintages and a good amount of the great novels of past centuries elaborate on this point. No person is infallible, and the isolated autarkist or his offspring/successor is likely to end as a tyrant. If we take that as an empirically sustained statement, it is in vain that one hopes that 'strong, undemocratic leadership' will be the solution when deep and urgent restructuring is called for. For those of my readers who think that my fear of dictators is uncalled for, I want to refer again to the interesting studies on the growing use of IT. China is already using the internet and its social media to control its citizens from birth till death, rating them on obedience on a daily basis: a bad score means one may lose permission to secure a university education, or even better housing conditions, and so on. Studies in the West speak openly of the emergence of surveillance capitalism (Zuboff 2019): commercial groups and intelligence services know almost 24/7 what you are up to when you have a PC, a mobile phone or any other sophisticated device. Finally, the appearance of more, successful dystopic novels (Hannah Yanagihara's latest [2022]; Powers 2018, etc.), films and TV series seems to reflect a growing concern about post humanist IT-driven ways of living.

- *A humanistic/democratic alternative?* If we are dissatisfied with either of these two alternatives, we had better work on the old humanism and bring it up to date.

The Approach of Amartya Sen

The questions here read: is there another road for tackling these major problems than the irrational and rather aggressive anti-humanist, anti-democratic answer of present-day populists and so-called 'strong leaders' on the one hand? Or than the hopelessness of outrage on the other hand? I will focus on Amartya Sen's work to illustrate this point.

Nobel laureate, economist–philosopher Sen had a lifelong debate with John Rawls, the philosopher of law who can be considered to be one of the main heirs of the humanist–Enlightenment philosophy of law and political thinking after the Second World War. It is no exaggeration to state that a generation of politicians has been trained in the West with Rawls' views as the beacon for choices about democracy, law, justice and so on. Sen published his own substantial alternative to Rawls' famous *A Theory of Justice* in a book that went under the title *The Idea of Justice* (Sen 2009). In his work Rawls had tried to point out what stand out in the philosophical literature as the basic principles of politics and legal thinking for a just and democratic state. Sen rightly points out that, much like forerunners such as major Enlightenment thinkers (Locke, Kant), this very influential theory is loyal to and in fact entrenched in the 'transcendental' thinking that is so typical of the western tradition. Summing up one could say the following reasoning obtains always, whichever empirical historical reality is referred to:

- The thinker looks at the texts at hand, believing them to be the best of human knowledge; these texts express the kernel of 'human ways' (on politics, on justice, on…).
- By scrupulous logical analysis the necessary, and indeed conditional, principles of good behaviour can be reached (or indeed deduced) through and in the texts.
- Adaptations of rules and prescriptions can further be discussed on the basis of the premises (principles) as long as they are consistent with the generally accepted and textually available principles.

Sen calls this a 'transcendental' view, and that is precisely what it is: the premises or deep-seated principles hold an authority that is fundamentally unquestionable. Indeed, refusing these premises will be qualified as a lack of reason, civilization or (worse, since it can hardly ever be subjected to any sort of scrutiny) so-called common sense. The research by Nader, Graeber and Wengrow, and others offers a painstaking analysis of the ways other intuitions, or critiques that are not formulated in terms of the 'given' premises, are dismissed as ill-founded, not understandable, backwards, etc. This type of unwillingness to hear 'unusual', 'other' or 'culturally different'

136 • Humanism Revisited

perspectives has a long lineage: it may in fact have started with Plato's reactions to the sophists. The latter were not considered 'wrong' in their ways of thinking because their arguments would not solve the problems people confronted. On the contrary, their arguments did manage to reconcile different opinions and values through pragmatic compromises. But the Socratic dialogues show time and again how these 'gibberish' sophists could not reason well in the game of 'pure theoretical knowledge', otherwise known as consistent debating or as 'purely logical rhetoric' (see argumentation philosophies such as Perelman and Olbrechts-Tyteca 1957; and Toulmin 1990). After the fall of Athens and, later, the collapse of the Roman Empire, Christian theologians promoted Plato's logical-consistency format to become the ultimate form of high western thinking – exchanging, in my understanding, his vague idealism for the religious transcendental tenets of the religions of the book.

When reading Sen's remarks on Rawls' tremendously dominant political philosophy, I am always reminded of this minority criticism of rhetoricians like Perelman and Olbrechts-Tyteca and Toulmin on mainstream thinking in the West. Toulmin (1990) is probably the most explicit on this point. He develops the view that one can find two main streams of philosophical thought throughout western history: pragmatic questioning and negotiating in view of temporary but satisfactory compromises on the one hand; and the stream of 'system builders', who went for internally consistent, all-encompassing systems where decisions and arguments are only deemed relevant if they comply with the indubitable premises of that particular system. The sophists, the humanists and the 'empirical epistemologists' (of the 1960s and later) all belong to the first stream, but the dominant philosophers will be found in the second and by far the most prolific group. The latter type of philosophy can ostensibly be used by political system builders, to be sure: durable structures, a state or empire building, the huge and often complex legal and juridical constructions – all of these invest in context-free textual reasoning, where local and pragmatic compromise are rather more distrusted than welcomed by the interlocutors. Not surprisingly, Sen offers examples from other than western traditions where problem solving in negotiation and with respect for local values and compromise are the rule, rather than the exception.

In the first place, Sen sums up the disadvantages of the transcendental view (2009: chapter 4):

• It avoids or frustrates the search for comparative perspectives: the transcendentalist assumes that his premises can be attributed a universal status (precisely because they have this transcendental quality of the religions of the book, with their unique God for all).

- It focuses on the institutional logic of a reasoning instead of broader, but often more varied, social realizations.
- It forgets about, or is blind to, the effects of the choice or rule on other regions of the world: the latter are believed to join the ranks, as it were, once they come to understand the logic and the contents of the transcendental proposal.
- It yields a mentality of isolationism in a world of inequality (the present-day identity tendencies may come to mind here).
- It draws on 'reason' in a rather narrow way: most often the interests of the institute (especially the state or the empire) supersede those of particular social or cultural groups. In an unequal world this will come at the cost of the latter groups, even in a structural way.

The alternative Sen proposes – dubbed social choice theory – goes back to mathematicians such as the Marquis de Condorcet in the Age of Enlightenment. These thinkers were trying to find a way to reach the 'aggregation over individual judgements of a group of persons' (Sen 2009: 91). This will be preferred to the transcendental view, even at the cost of eventual inconsistencies. In that way, instead of obedience to a superior or supposedly transcendental principle, this approach aims at 'making social decision procedures more informationally sensitive' (Ibid.: 93). In practice, knowledge about the choices and decisions will not be 'deduced logically' from the transcendental premises. Rather, it will be built up through painstaking empirical research, leading to comparative work on the decisions, values and preferences of the persons and groups concerned. In that sense, it is much more an approach that rests on the practice of democracies and the concrete knowledge and habits of individuals and groups. Of course, this need not exclude the possibility that some of the transcendental rules and choices might prove to be adequate or the best option. They might prove to be acceptable after open and respectful negotiations, listening to the arguments of all involved to begin with. But the a priori superiority of transcendental decisions and the context-independent superiority of consistency rules are not the guiding principles.

In a succeeding paragraph Sen continues his reasoning by remarking that the present predicament of the world is one of (factual) interdependence of interests. This is, as the reader will have noticed, very much in line with the perception that I have of the world and the possible survival of humanity today: I judge that global interdependence is a factual condition for survival. A consequence of this condition is that the interests of humanity should first of all be understood in this context of interdependence, and not be restricted to (some) individuals. Sen mentions that this condition of interdependence works in such a way that injustices spread more easily and more widely than ever. Isolationism, or the island mentality, is becoming more untenable because of this factual interdependence. Hence, apodictic, transcendental premises eventually become unworkable or out of date.

Having mentioned all this, it is clear that Sen works from a different perspective on humanity than the hierarchical, diversely 'heteronomic' view from which the transcendental reasoning emanated. This perspective is expressed in his comparative theory of human capabilities, which I will dwell upon in a later section of this chapter. But before I go there, I need to address a further issue: we, westerners, developed scientific knowledge that pictures reality in a rather dependable way. The 'others' did not do that, it is believed. I claim this statement is unqualified, too blunt. Let me explain.

Scientific research is producing the best, the most dependable sort of knowledge humans have been capable of so far. I subscribe to that statement fully, including the proviso that I reject relativism. As a method of investigation this type of research is very strong, provided fallibilism remains one of its pillars. As I indicated before, the methodical approach is applied within the perspective (the frame of reference, the intuitive window, the ontology) which is chosen. Changing the perspective may be induced by scientific insights, but is external to the method as such. Also, scientific research is subjected to contextual conditions: in each era – depending on the type and source of funding available, the political wind that is blowing, and the value and taste preferences of the era – inspirations will differ.

So yes, we have the best method to develop knowledge – but no, this does not guarantee that the best knowledge will automatically be generated.

An illustration is provided by what I can only call the scandal we are living today.

The Scandal or the Delusion

When life goes on we humans live it from day to day, mostly without putting our daily experiences into a long-term perspective. When something out of the ordinary happens, we revert to that perspective, though. Over the past few years, we became aware of our vulnerability as a species with the COVID pandemic. When COVID-19 was approaching what looked like a slowdown and maybe even a decisive final curve (beginning 2022) the war in Ukraine started, showing horrendous brutality. It demonstrated once again our vulnerability. In the latter case this was caused by the dependence on (enslavement to?) fossil fuels for the economy, but also for comfort in our private life. When looking at the sequence of events in the long term we can imagine the following: historians of 2050 and later will, with high probability, describe political decisions made in 2022 as

Reset or the Extinction of *Homo Sapiens?* • 139

astonishingly scandalous or maybe delusionary. I suggest some lines of interpretation.

For one thing, it was of course reckless for a nuclear power (like Russia) to wage war on a neighbour and (re)conquer their land, thus risking other nuclear powers' retaliation. In military terms we see for the moment (2022–23) that western forces supply enormous weaponry to the Ukrainian army in order to fight their battle against Russia, which retaliates with massive destruction of civilian targets. That is one scandal.

At a more general level, something else is at stake. For half a century scientists have produced evidence that the unlimited extractivism is increasingly and rapidly extinguishing thousands of species, and is causing a deep, structural disharmony in the ecological environment and in the climate system of the earth. For at least a couple of decades evidence is amassed which shows that humanity is putting its own survival in danger.

Throughout that period, lobbies from the fossil-energy corporations knowingly launched so-called 'alternative-fact' research groups, where a small number of dissidents essentially spread doubts about the work of serious and respectable colleagues such as the scientists of the IPCC. Populist politicians have been spreading outright lies and have launched a dangerous slogan like 'alternative facts' (in the name of freedom, above all) without being punished for it. Lobby groups for the fossil-energy giants intervene in political decision making (like the COP meetings) in order to defend their interests, at the expense of very poor populations in the Global South. Instead of putting such lobbyists and their sponsors on trial in the name of the global interests of humanity and the earth, the scandal a scientifically informed West is living gives way to absolutely surreal events:

- A little (in population terms) gas- and oil-rich country like Norway, with a so-called centre-left government, decides, after signing the Glasgow COP agreement (with 2050 as a binding deadline for a drastic reduction of fossil-fuel dependency), to grant permissions for exploiting new gas and oilfields in the Nordic area for the post-2070 era. In other words: private interest goes beyond global commitment, even when enormous casualties will result from further climate warming.
- When Russia attacked Ukraine in 2022, the climate issue was suddenly removed to the background in political debates and in the media. The rise of the price of fossil fuels competed for months with news about atrocities committed by the invader. An inimical atmosphere from the time of NATO–USSR opposition (the so-called Cold War era) seems to be back. The link with the climate problem should manifestly be stressed here: even a short-sighted thinker and decision maker should jump at the 'opportunity' of the Russia–Ukraine war and develop necessary and

feasible projects enabling the transition from the harmful dependence on fossil energy towards policies stressing renewable energy. What we got instead was that the old political scenarios (President Vladimir Putin for Russia, Chancellor Olaf Scholz for Germany and President Joe Biden for the USA) continued to play their role to safeguard fossil fuels almost completely. Putin blackmailed the EU by manipulating the gas-supply system. But when US President Biden came over to the EU and NATO headquarters in Brussels, Belgium in 2022, he actually negotiated two contracts: most European partners promised a substantial build-up of investment in their military capacity and budget under the NATO umbrella, and the president promised to send another and rather substantial cargo of fossil fuels to the EU. The military build-up is unproductive money wasted: weapons are costing a lot, but those budgets do not yield more prosperity or wellbeing for the population. The fossil fuel that President Biden promised will notably be gas from the ecologically devastating fracking industry. There was not one word mentioned about a mutual agreement (in the COPs) to use this war as an opportunity to step up the necessary transition towards non-fossil energies. One need not have evil thoughts to see this type of political approach as 'old school': the fossil-fuel corporations won and the sustainability thinkers; the climate researchers; and, in fact, everybody who cares for humanity as a whole and for nature lost once more.

Looking back at this episode historians in a generation from now (say, 2050) will vary in appreciation, I guess: some will speak about a missed opportunity and some will conclude that this 2020s politics was a scandal. Knowingly and willingly our political leaders (in the democratic countries and in the authoritarian parts of the world alike) took decisions that in all certainty caused further disruption, pain and disasters in a world in climatic crisis.

When trying to stage legal actions against this kind of anti-human and anti-nature ruling (the way the Climate Case organizations in several countries have been doing, with relative success), states and governments are sometimes convicted for negligence vis-à-vis the interests of their citizenry. But, more often than not, the actual changes in policy that would have to follow are postponed. For example, Guatemala was convicted but nothing happened afterwards, and Belgium managed to suspend any actual court order by the Climate Case groups because of the difficult and opaque legal–political structure of the country (Flemish and Walloon courts advertised their incompetence in a matter that transcends this linguistic–political division). Moreover, jurisdiction can convict a person or institution for a particular crime but the procedure will have to be repeated when and if even the slightest 'adaptation' of this bad behaviour is put in place. Jurisdiction does not detail what 'better' behaviour should look like: it can only point to wrong or criminal acts. Of course, when taking the point of view of the generations of the future (as Yanagihara does in her dystopian novel *To Paradise* [2022]) and looking back from there to the present time, there is no other word that comes to mind, I guess, than 'scandalous'. Short-term power thinking, short-term gain for the elites that be who reign supreme: majorities in both well-to-do and poor countries are clearly the victims of these elites since the continuing crises (first and foremost, the climate crisis) will progressively hit the so-called 99 per cent hardest. Meanwhile, their voices are not heard since the old political and economic doctrines exclude them.

'The Western Value of Progress Is the Only Guarantee'

This is another assumption, or credo, of which an updated humanism needs to be freed. After a life of questioning, the anthropologist David Graeber passed away at a rather young age, leaving us with an important book he had composed for over a decade together with archaeologist David Wengrow (Graeber and Wengrow 2021). I return to this book one last time, since it addresses for the first time in a synthetic and thorough way the problematic nature of the deep-seated conviction of humanism and of Mediterranean religions that I focus on in this section.

The title of the book is *The Dawn of Everything*. This can be understood as a very ambitious title, or it can be appreciated as a hint at the content or the basic question of the book: where did it start? What could be the origin of this belief in economic progress that has been so strong and often reckless for at least five centuries in the West? I am not a historian, so I leave the first question to my colleagues in that discipline. The second question is intriguing and promising. How does it work out?

The authors combine anthropological interest in other cultures (Graeber) with an alertness to recent findings in archaeology (Wengrow). The dominant view in the West was and is that human history evolves according to fixed order in time: first humans are hunter-gatherers, then agriculturalists, then industrial agents and finally post-industrial consumers, helped by post-human agents (cf. the work of the popularizing historian Yuval Harari [2015]). This history of humanity has a clear, unidirectional line of development because every step of it is understood as the next one in a linear continuity of progress. Time is conceived as an arrow, and only Westerners 'are sitting on the arrowhead' in this view. The economic extraction of resources from nature, and the ever-more prominent submission of peoples and creatures on earth to the dominance of technology at the service of the so-called free market, is the driving force of this development.

142 • Humanism Revisited

Along with the dualistic ontology contained in it (humans versus nature: Descola 2005) this particular idea of progress is framed as an essential feature of the development of humanity. History led, especially or more impactfully since the Enlightenment, to the massive extinction and/or enslaving of other cultural traditions, as an obvious part of the 'cultivation' of every possible piece of land and the extraction of any profitable resources from the soil all over the planet. The notion of progress, which was thought to be uniquely instantiated by the westerners and their view of humanity and nature, was an essential ideological element in this endeavour. Over the period of a decade Graeber and Wengrow gathered evidence which unsettles or might even smash this self-image of the Westerner in a rather convincing way:

- New archaeological work proves beyond doubt that large constructions, even urban concentrations, were erected by large groups or communities long before the agricultural era started. This happened in different parts of the world, and apparently did not always yield the presupposed hierarchical political structure which is deemed a necessary step 'forward' in the received view. New findings keep emerging, and the new means to determine with high precision the historical time of their appearance as well as the contextual conditions of the builders (gatherers rather than peasants, large groups and so on) are rightly summed up as refutation of this old view. Indeed, according to that view such constructions could only be realized by hierarchical agricultural communities. *Quod non.*
- The authors combine these findings with a thorough analysis of the historical data on the 'first contact period' between Europeans and many 'native communities'. Graeber and Wengrow present stacks of evidence of what the 'others' thought and often said about us, the self-declared representatives of progress.

The authors list printed sources, and relate public discussions and statements by missionaries and travellers. All of this material shows that the so-called 'primitives' whom the travellers/conquerors met in the 'new countries' were able thinkers and discussants, who developed and sometimes expressed their critical opinions on the 'Whites' they encountered. Often they met the colonialists at first with goodwill or curiosity. In a second reaction they sometimes expressed surprise, disapproval or even horror about particular aspects of the so-called civilized foreigners they found on their territory: some stated that the inequality among people (and more particularly between men and women) was disapproved of, and they often rejected the way the White conquerors saw natural phenomena (including humans). Although Graeber and Wengrow do not refer to the ontological discussion mentioned by me in this book, native interlocutors reasoned much more in the holistic frame of reference where 'interconnectedness' (as we call it today) is considered to be an intrinsic and certainly a cautious way of thinking about 'nature'. The dualism and the ensuing emphasis on the private ownership of natural resources, and finally the grabbing and 'cultivation' of land for agriculture and the quasi-industrial exploitation of domestic animals was quite generally met with disbelief and disapproval. It was when the new settlers started taking land, cutting down forests and chasing native people (often killing them off) that the latter decided to turn against the White conquerors in so-called tribal wars. Meanwhile, discussions had taken place between natives and invaders where the former expressed their unwillingness to either form a part of or be integrated in a hierarchical political structure (county, state or whatever). Sometimes they conversed with the settlers and their army. Quite often they abandoned that approach, or they decided against it after negative experiences. In other words, we seem to have sources for a couple of centuries that relate the fact that native peoples in the Americas and elsewhere had discussed political organizational formats and decided against the structural hierarchical society that the Christian European colonists thought to be the most progressive or 'civilized' in the world, and destined to become the rule for the whole of humanity. Graeber and Wengrow show how some of this was known in political circles in Europe at the time – or at least in certain circles, since some of the native leaders were brought over to 'tour' royal courts at the time (e.g. in France, where Montaigne saw some). It was only when the economic importance for the European homelands (gold; silver and other materials; cheap labour: see also Ghosh 2022 for Ambon (in present-day Indonesia); Needham 1965–2000 for China) and later for the settlers in the new colonial context (agricultural land, grazing land, oil, etc.) became clear that the knowledge of and the appreciation for the native point of view changed dramatically: the native peoples were then rather systematically framed as 'primitive', heathen (and thus morally and religiously wrong), a mere workforce (as slaves) and utterly politically irrelevant or harmful. With the onset of industrialization and the capitalist conquest, helped by early science and technology (with James Watt's steam engine as the historic marker), the image of the non-European as a right-less primitive was firmly established. Even after the abolition of slavery in Europe and America, discrimination against people from other parts of the world was not put to an end (Hari 2023): without doubt this attitude is still very much around in the western worldview today, which accounts for the fact that even humanism nowadays has not deeply adopted an inclusive approach yet (see the NE part of my call for a reset of humanism).

The contributions of archaeologist Wengrow to the discussion further enlarge the scope on human history, and particularly on the general notion of human progress (from hunter-gatherers, through agriculturalists, to industrial so-called 'free market' societies with democracy). According to

144 • Humanism Revisited

that view progress can be measured by the development in the later phases (starting with agricultural societies, in fact) of hierarchical societal settlements with high authorities (kings, etc.) creating large architectural constructions: temples, palaces. Indeed, dating was and is largely thought of along these lines: agriculture started some ten thousand years ago, the story has it, which is witnessed by the emergence of power hierarchies and the building of large constructions. However, over the past two decades, Wengrow emphasizes, large built complexes have been found that date from two thousand years and more prior to the advent of this agriculture. This means that, for reasons unknown to us, great masses have been working for a long time in different parts of the world erecting large constructions without a permanent elite in a hierarchical organization of society (a high priest, king and so on). Put differently: for whatever reasons, people have been choosing for, thinking through and developing these complexes for a certain amount of time, but later left the place and the collaborative large groups to return to their normal way of life (as gatherers).

Also, particular cultural groups have been documented recently which switch between egalitarian small societies (e.g. during summer time) to take on the form of hierarchically organized forager groups for a certain part of the year, and return to the small 'horizontal' groups in a following period of their existence. Again, the linear notion of progress which Europeans (and Whites in general) projected onto a presumably necessary development of humanity from 'primitive' to property-driven, civilized formats (such as the European–western society of the past three centuries) is a largely ideological construction that is refuted by these recent findings. It was instrumental in the colonization of the 'others', of course, just as it still works wonders in allowing for economic globalization – both in the version of capitalist worldwide expansion and in that of the Chinese new Silk Road (Frankopan 2018).

So, I am grateful for the painstaking research of Graeber and Wengrow that resulted in this remarkable study, as I am for Ghosh's (2022) treatise and some others that are now being published. Still, I have the feeling that the work is far from finished. For one thing, the authors cited work in rather isolated subgroups: Graeber and Wengrow in anthropology, archaeology and history; Ghosh in colonial history; Needham on the history of China; and so on. Other studies have emerged, however, or have been around for some time, and they need to be integrated so that a strong and convincing frame of reference can be put in place. I think of the historical work in the systems approach of Immanuel Wallerstein and others (see, e.g., Stuurman 2010). But there is more: the findings of such important 'loners' as anthropologist Laura Nader (2015) should also be integrated. In the 2015 volume *What the Rest Think of the West* Nader presents a series of texts from quite varying geographical origins (Asia, North Africa, etc.) by people who have written short or long essays on their experiences when coming into contact with Europeans. The texts date from the Middle Ages in Europe right up to the past century. Diplomats, salesmen and other visitors to Europe put their impressions on the ways of the Europeans down on paper, and Nader presented a selection of this material to show that 'others' did have remarkable sharp opinions and ideas about what Europeans thought to be the unique and presumably most developed way of 'being human', which only they could exemplify. The 'mirror' that is thus held up is, at the very least, intriguing.

In all fairness, yet other sources should be added: the path-breaking studies on temporality in different cultures and their encounters by Fabian (1984) and Wolf (1981), among other work. So, although I applaud Graeber and Wengrow's work I advocate a broader scope still, that will recognize and integrate other and more or less isolated, related studies. This will certainly make the perspective even more convincing. However, these inclusions cannot diminish the value of the study by Graeber and Wengrow – especially with regard to the point where an ideological, and indeed partly fictitious, notion of progress is shown to be not a scientific truth but rather a conceptual means of domination and discrimination to the benefit of Eurocentrism, White supremacism and the like.

The point is that a notion of material economic progress has been used ideologically in such a way that it justified or excused the conquest and the brutal elimination or 'civilization' of people all over the planet. Also, it allowed for a war against nature and the extraction of anything of value in the frame of market economics. The resulting disasters in terms of human pain, destruction of species and climate warming were at best seen as 'collateral damage', which would be compensated for in due time by the global adoption of western superior science and technology. Does this critique imply that the development of the past few centuries did not result in anything good, which could hence be discussed as 'progress'? I do not think so, but that is a question that needs to be addressed through thorough empirical research: what is better, and in terms of which goals and values do we qualify it as such? It is clear that adopting or rejecting the interdependence perspective will yield different answers. The critical point for the humanist, however, is that a conqueror's ideology has been successful in progressively claiming these intellectual developments to put them to work to a large extent for the benefit of the new elite. The capitalist, entrepreneurial attitude replaced the 'free men research club' of Isaac Newton's time. Along the way, the so-called humanist prerogatives of freedom, individualism and human interest have been used as justification to stay on this extractivist path.

146 • Humanism Revisited

Similar remarks can be made in other domains: of course, there are more products for sale in large parts of the world than ever before. But how should one evaluate this outcome as 'progress'? Does it make people happier, more equal (Piketty 2019) or more in harmony with the rest of reality? What are the most sensible standards against which progress is to be measured beyond the mere amount of products sold, so beloved of the market economists? How can we organize an inclusive negotiation process to determine what will be called 'progress', beyond mere economic and material surplus? Is Sen's capability theory (see below) the best tool here? These questions are fundamental for anybody who aims at living with other humans and with other natural phenomena in a harmonious, non-destructive way, preferably within a long-term perspective. In my proposal, the notion of progress that will be used should then be thought of and phrased from the NEED perspective. It will most probably be a very different one from the economistic version we have been brainwashed with, at the expense of other cultures and of 'nature'. So yes, the horizon is open to new perspectives: less is more maybe, or quality instead of quantity, and certainly also inclusion instead of ever-more exclusion and war.

How Could Humanity Accomplish a Deep Shift towards NEED Humanism?

It is time to address the million-dollar question and tell my readers something on the 'how' question. Maybe I should even try to formulate some concrete ways out, make some dreams somewhat more tangible. For this I return first to Sen. He developed, together with Martha Nussbaum, some more practical philosophical schemes for education, interaction and societal decisions that might allow us to avoid falling into the abyss. That is how I appreciate their approach.

The humanist–Enlightenment perspective presupposes that individual and rationalist stands can in themselves be seen as the universal basis for moral and political life. Rawls' theory is the most successful contemporary emanation from that long line of thinking. However, in the course of history it became clear that universalism would not obtain automatically, or even at all, or the presuppositions proved to be too particularistic (see, e.g., Sen 2009, Nussbaum 2011):

- Equality between humans of different gender, and between people with varying (dis)abilities, race or religious-/life-stance orientation did not obtain automatically. On the contrary, discriminations on all these lines persist in a variety of ways (depending on local history, etc.).

Reset or the Extinction of *Homo Sapiens?* • 147

- Nationalism grew together with the ongoing structuring of the planet into states over the past few centuries. Nationalism proved to be a stubborn, irrational force readily used to motivate wars and massive impoverishment, or even annihilation of populations. The new format of cultural identity (as in Generation Identität, for example, one of Europe's fastest-growing far-right movements: Höhne and Meireis 2020) proves to overlap with despicable forms of nationalism from the Second World War.
- The earth is not included in most universalist views: other species and the climate system are still looked upon as sources to be mined, managed or destroyed at will. Rationality is an impactful way of interacting within a particular view of reality. But, as with anything humans might say, what is universal about it? People think and produce mental representations of reality, themselves included, at a superficial level. When we examine the process and the specific representations, it appears that we can only point to particular procedures and to local and temporary products of thinking. To illustrate this point: associative reasoning and analogical thinking will be found in many communities and thinking traditions around the world (Descola 2005), but the way these forms of reasoning are used differs – one can apply them to everything, or only to a small selection of phenomena which are seen as relevant or important in a particular tradition. Moreover, the intuitions or presuppositions used will differ from one tradition to the next: the holists will claim that the systematic erasure of context in the textual tradition of the West will automatically lead to destruction of the environment, and finally of human beings (see, e.g., Boyd 1974; Farella 1984;and others for Native American traditions, and many sources on eastern ways of thinking about reality). Western thinkers, on the other hand, will emphasize that selective and context-free thinking is a precondition for rationality and has been shown to be very powerful.

Arguments on both sides can be mentioned: the detached and context-free thinking of the West has been very impactful indeed in science and in economic expansion. On the other hand, holists will retort, it is precisely the neglect of the encompassing whole (and hence of the context) which has led humanity and the earth towards the dangerous state of affairs we are confronted with today. Also, rationality in the Western sense is not factually universal, and the claim that it should be promoted to become universal (through education) is mostly blind to the recklessness I discussed before. The war cry is 'Go and change at your will. Solve any problems that emerge during this conquest by using the same approach more powerfully'. Waste, economically and politically 'useless' or counterproductive elements

148 • Humanism Revisited

or a 'stubborn' refusal to fit in with this approach to reality by nature or by other cultural traditions can be handled by pushing them aside or eventually destroying them. Such principles are not reached through rational thinking, however; rather, rationality is historically almost always at the service of such points of departure. We have slowly come to realize that all peoples clearly think and use reason, but the content and the procedures differ. The claim that rationality as it is understood in the West (detached from context, mostly atemporal, selecting only what is manageable) should be used universally did not necessarily prove convincing for 'others'. The intuitions on which the western use of rationality is applied are not universally shared – e.g. the almost exclusive focus on material aspects of reality, the perspective of extractivism, the textuality approach and so on. When discussing such issues, morality – even ontology – is involved: conscious and conscientious deliberation should precede rational thinking and acting.

All of this argumentation makes clear that Sen's and Nussbaum's claim that value decisions are unavoidable and should be made before engaging in 'rational thinking' holds water. They started developing this argumentation with a critical evaluation of the detached, rationalistic views on freedom, the good society and human development which accompanied the first humanist proposals.

Sen and Nussbaum developed what is commonly known as the capability theory. This holds that, together with the juridical frame of human rights, one needs to recognize that ethical choices should be included. Capability refers to substantial freedom, that is to say the guarantee that everybody will at all times have the chance to choose how to handle matters that arise. It is seen as a precondition for human interaction: recognizing capabilities in all humans implies that everybody should therefore be treated with equal respect. Opting for this approach entails abandoning the doctrine of meritocracy in the free-market ideology: nobody deserves more than any other fellow human being because of her particular ability. Also, biological or genetic dispositions cannot suffice as an argument for treating people differently. Both can be seen as resources, means to help realize potential; however, they are only means at the service of a goal. The goal is of a moral nature and is captured in the aim to guarantee chances to all. Nussbaum (2011) lists the capabilities one should opt for in ten points.

1. Life. Guaranteeing to all the right to live, and live in such a way that one's life will have a normal timespan.
2. Physical health. Allow all people to live in healthy conditions. I add here: with climate warming this moral principle takes on a new dimension.

Reset or the Extinction of *Homo Sapiens?* • 149

3. Physical immunity. One should be allowed to live free of violence and able to freely choose sexual and gender orientation.
4. Guarantee the chances to develop observation through the senses, imagination and thought.
5. Feelings. One should be allowed to build human relations of love, gratitude, desire and justified anger.
6. Practical reason. Everyone should be given the opportunity to evaluate life, with freedom of conscience, and be allowed to reason on what could be good.
7. Social ties. Every human being should be allowed to develop firm and responsible ties with others, in a process of free choice.
8. Other biological species. This is added by Nussbaum to expand the old humanist frame and include other species. In my proposal this should be expanded further to encompass the earth (climate, all natural resources and so on).
9. Play. Humans should be allowed to play; or, in my words, recognize human creativity.
10. Built environments. This capability refers to human capacities to build material things and contexts, but also political constellations.

On all ten rubrics a general engagement for justice and equal opportunities for all has priority over any juridical or ideological value or norm. When looking at the real world, such an emphasis on human capabilities goes deeper than, and needs to be addressed prior to, such values as Gross National Product or so-called basic human needs. In that sense it is an addition to, and in fact a foundational theory for, the juridical frame that can be found in the Universal Declaration of Human Rights, Nussbaum claims. One of the ways the capability approach could work is by determining basic and universally agreed-upon moral values first, prior to juridical regulating. Hence, when conflicts arise in practice one can return to the basic moral choices and avoid sophisticated and often discriminatory juridical battles. These have proved discriminatory in historical practice because, more often than not, power and financial differences have been shown to be more important in real life than the guarantee that all will be equal under the law: in actual fact it often takes good and expensive lawyers to fight for your rights.

Within the non-Eurocentric perspective that I advocate, I would like to summarize this important reasoning in words inspired by Native American medicine man Rolling Thunder: 'Orientals [and Native Americans] have always thought more of responsibilities and duties than of freedoms and rights. They don't share our concept of the self as an isolated, individual identity: Westerners dislike a sense of duty...' (Boyd 1974: 152). When

150 • Humanism Revisited

looking at the approach of Sen and Nussbaum from this comparative view on cultural diversity, the focus on individual freedom (in the old humanists) and social contract and human rights (in Enlightenment elaborations right up to Rawls' synthesis) invests in the legal and rights conceptualizations and shies away from deep engagement in responsibilities. In the light of the disasters that this approach caused or helped to justify over the years, that seems to be a constant: a (moral) discussion on responsibility and hence negotiation on just, fair and shared values and choices is avoided while expensive and sophisticated battles over rights and freedoms (of choice, but also of private ownership) in courts can lead to inequality (Piketty 2019), and to the endless postponement of the courageous choices we need to make as responsible and inclusive humans-interconnected-with-nature.

But how does one work with the capability approach in order to change the existing state of affairs now that the danger of self-annihilation seems imminent, inequality is growing, etc.? As indicated by Nussbaum, and even more explicitly in Sen's profound reaction to Rawls' way of thinking, human beings need to consciously and conscientiously consider all aspects of a cognitive, moral and political nature in their interaction with other humans and with the natural context:

A. Cognitive Aspects

It is clear that processes of gathering and building knowledge are embedded in long traditions, offering particular intuitive routes to approach and understand reality. This point yields a broad scope of possible investigations: what is the role of linguistic deep structures, for example? Also, as mentioned before, what is the role of particular contexts, both natural and socio-political?

On the first question, one perceives and constructs reality differently when the deep structure of one's language consists of nouns (noun phrases, etc.) and verbs (verb phrases, etc.) as in the Indo-European languages, or (almost) exclusively of verb forms (as in classical Chinese, the Athapaskan languages, etc.). It looks reasonable to see the development of geometry and the so-called 'geometrization of the physical world' (Einstein 1949 [Schilpp 1969]) occur in the former type of deep structure, whereas the latter will produce a processual worldview and eventually a type of algebra rather than geometry (as in the Chinese proto-scientific development: Needham 1965). The process view on reality will in the first case be exceptional, a minority point of view (Whitehead 1929; Prigogine 1969; and Prigogine and Stengers 1984).

At a possibly even more foundational level, I pointed to the role of (mostly subconscious) intuitions and their impact on the relevance or (un)likelihood of particular questions and research lines. Thus, the holist looks for signs and messages from plants and animals in the environment, and the dualist rejects such a line of thinking in principle. Humans are interconnected with all other phenomena (right up to winds, mountains or the sun in several cultural traditions) in the holist view and relevant, sustainable knowledge will typically integrate a particularly wide range of 'data' in order to render a dependable overview of the harmony between everything and the lines of impact between all phenomena in their networks of interconnectedness. This view on reality contrasts in a deep sense with the reductionist and context-free intuition of the West, which starts from the dualistic view on nature: here, humans are the only creatures with consciousness and hence with the capacity to understand a supposedly inert nature.

Depending on these deep-seated intuitive concepts particular data will be considered relevant, based on facts, or not. Also, particular lines of questioning will be allowed for or not. In the case of western history, a mechanistic, time-independent focus became predominant to the exclusion of other perspectives. In past centuries, the forceful impact of systematic, scientific knowledge working from this perspective showed a preference for reductionist intuition: split up the whole into smaller parts, which are supposedly more convenient to study, and then make the sum of all partial knowledge items in order to understand the whole. Along the way, process or 'real' qualitative change was thus abstracted from the perspective on reality (Prigogine 1969). It is only with the coming to prominence of thermodynamics and of the biological sciences that this intuitive stand is slowly and hesitantly being criticized.

The depth of the impact of such intuitions is illustrated by the analysis of mathematical thinking in the approach of the brilliant topologist René Thom: in his attempt to overcome the 'time-independent' and 'geometricized' static view on reality he initiated the development of a dynamic topology and applied it in biological and linguistic research (Thom 1987).

It is clear that at this general level of cognitive processes rather little is known about alternative views on reality from other cultures, let alone implemented in the educational and research programmes so far. I mean, it is possible and would be mind-opening to at least integrate knowledge about different intuitions, and the accompanying ways of reasoning on the basis of them, into the schooling of every new generation. For example, instead of installing the textuality tradition and its exclusive perspective on nature and humans through worldwide schooling, it is possible to teach a diversity of human views and of learning types at every level of schooling. Storytelling, the exploration of cultural practices, ceremonial activities and the development and use of local technology are all means to educate subjects in a wide variety of cultures. That way, knowledge would be genuinely 'situated' (Lave and Scribner 1986): it can be identified as rooted and, initially in principle, local. Realizing this and using these means in education might prevent educationalists from making imperialistic claims on behalf of their particular line of knowledge, on the presumed basis of its efficient or convincing way of solving problems. I refer to the way schooling has been implemented worldwide as the only valid way for education for past generations. To be sure, western scientific knowledge has been shown to be very successful in mapping certain aspects of natural phenomena. But by forcing it on others as the only civilized way of education it disqualifies other perspectives on reality apart from the mechanistic, dualistic perspective referred to. It now appears this approach is too narrow in scope to depend on for the survival of humanity. We need to grow out of this narrowness but, for example, the deafness of western powers to the cries for help from drowning Pacific Islanders, the first victims of climate change, shows that opening our minds proves difficult. It remains to be seen in whether and when a substantial overhaul will happen in order to meet long-term goals such as sustainability for human survival. For now, the short-term benefits of small elites persist and drive the course of science, technology and of education through schooling, regardless of durability. Morality is absent from knowledge proper, it seems. Still, modest and small-scale alternative developments should be mentioned more often. One such development I want to emphasize one more time can be found within the domain of mathematics education: 'ethnomathematics'. Mathematics may be the ultimate example of cognitive performance, understood to be context-independent by definition. In that sense it is the best illustration of the universally valid approach to formal reasoning, to the exclusion of all other alternatives. Secondly, it is held that this type of thinking was developed in the West, starting with Ancient Greece, and shaped as a specific form of deductive thinking. It followed from this view that all other forms of more or less formal reasoning around the world in games, in cosmologies or in technical devices either would have to be false and/or amateurish or would be seen as instances of the same universal tradition that had been successfully established by Westerners. Particularity does not obtain, and the universality of what constitutes mathematical knowledge is unquestionable, it is claimed. Following this line of thinking it was obvious for all in mathematics education that only what western mathematicians conceived as mathematics could be part of the curriculum, to the exclusion of all other forms of formal thinking.

In practice, the implementation of mathematical knowledge led to some interesting facts: orthodox Jewish students keep having difficulties with geometry (due to the linguistic deep structures of Hebrew?), some Native American populations are high performers while other communities show high dropout rates, pupils in western schools show a consistent gap between good and bad performers explained away by 'innate talent' or lack of it, and so on (see Pinxten 2016 for an overview). When, finally, some educationalists started to look at knowledge building and learning as 'situated' processes, it became possible to understand and explain these differences. Within a few years the subdiscipline of ethnomathematics was formed, emphasizing that learning processes for mathematical knowledge are indeed best understood by looking at the cognitive makeup, pre-school worldview and linguistic context of the children in the classrooms.

When the mental set-up of children coming into class is disregarded, chances are that many children will be unable to find the necessary utensils in their cultural background to provide cognitive bridges and reference concepts which can help them understand and recognize the new schoolish content. Ethnomathematicians started studying the out-of-school formal concepts and procedures in the particular linguistic and cultural environment of the child. The technologies developed and used successfully in many cultures – the architectural sophistication, the cosmologies, the formal aspects in rituals and in music, and so on – all these showed a huge variety of formal or abstract reasoning. Ethnomathematical educationalists would then go on and start organizing further learning processes on the basis of these local or culture-specific skills. This means that subjects of any other culture are no longer seen as 'blank' or inert in formal thinking, but rather as thinking subjects while being raised in more or less different perspectives on reality. The process of education can then turn around: no longer will the teacher erase the supposedly wrong or 'undeveloped' notions and practices of the pupils in order to plant the one true doctrine in their heads. Rather, the educational process will develop as a dynamic, interactive relationship where the local categories are taken to be valuable concepts and procedures on which more and other notions and views can possibly be grafted. Or, to put it differently, education resembles more a process of intercultural negotiation than one of assimilation. Thus, a holistic approach at the intuitive level is respected and can gain the same status as the dualistic view of the westerner, if not a better entry in the field of formal reasoning (Baker 2023). Or, the implicit rules of local games can form the basis of teaching, or the curriculum can be largely turned around to start with the pre-school knowledge of the pupils rather than the concepts of European mathematicians (Vandendriessche and Pinxten 2023).

When looking at the list of capabilities formulated by Nussbaum, it is clear that rubrics 4 and 6 are reconceptualized here: my point is, clearly, that the general perspective of the Sen–Nussbaum proposal is adopted, provided the particular content of each rubric is comparatively substantiated. Indeed, in contradistinction with the 'old' humanism and Enlightenment conviction, I take the stand that 'real emancipation' or 'giving chances' and

154 • Humanism Revisited

granting opportunities for the full deployment of sense, imagination and thought (4) and of practical reason (6) will be shaped differently by particular agents. Linguistic and contextual parameters will be different from one group to the next, and the moral choice Nussbaum so emphatically wants to promote will hinder emancipation if the values of these parameters are understood in the uncritical or even in the 'charitable' superiority attitude that westerners often radiate. Keeping in mind the deplorable history of assimilation and of outright genocide the Eurocentric approach represents (see, e.g., Fisher 2017), it will be obvious that a self-critical rendition of the humanistic project of life will have to deal with this ugly side of its past.

This critical rethinking and redirecting of content should be done for each and every rubric, along the general lines of the NEED view advocated above. This will make the theory of capabilities genuinely emancipating, I claim, and even allow for a concrete elaboration of pluriversality as mentioned in the NEED discussion. Indeed, when diversity is taken for granted and is evaluated as a bonus rather than a handicap, people are invited to seek out negotiated solutions for living together. Sen's proposals (Sen 2009, but also 2002) indicate that this is an almost unexplored avenue for political and economic problem resolution. With factual interdependence as a condition of life on earth, intercultural negotiation will have to become a necessary capability (rubric 10 in Nussbaum's list) to be promoted. Updated humanism will allow choices and opinions of diverging origin to be entered into respectable dialogues and negotiations, as was the case in several traditions throughout the world: the Leagues of Native Americans come to mind (Morgan 1852), but also several of the historical examples of local pluricultural societies (see, e.g., Sen 2009). Obviously, this change in attitude requires serious readjustments: learn to listen to each other, find ways to communicate and negotiate viable and sustainable solutions for survival problems in an atmosphere of mutual respect rather than discrimination, and allow for diversity rather than dogmatism (Hari 2023).

It is clear that each of the rubrics listed by Nussbaum will have to be thought through and explicitly rephrased in a similar way. In the second place, each will be reformulated in particular negotiations involving other cultures, but also other species and 'the earth'. It would be awkward or even pompous for me to do that job all by myself: the suggestions on ethnomathematics will have to suffice in pointing to such an overhaul.

B. Moral Issues

A central issue of morality for humanists seems to be that the individual is freeing himself from collectivism. The latter was, of course, the dominant socio-political structure of Christian institutions in Europe: people were defined by birth as belonging to one of the three strata, which determined in no uncertain way their rights, obligations and life-expectancy chances. Any claim to more freedom or any positioning outside of this socio-political classification was likely to be answered by violent and life-threatening reactions from the combined forces of clergy and nobility: women (and some men) could be identified as witches and eliminated without further regard for the qualities of such an individual. Free-roaming people were basically considered to be sinners, and artists and 'thinkers' (theologians) worked obediently within the worldview and at the service of the powers that were. In a pendulum movement, I would say, the Renaissance and humanism understandably stressed individual rights and individual capabilities against this crushing collectivism of the so-called well-ordered society. It is not surprising that rightist thinkers of the present era will somehow dream of a return to such a 'well-ordered society', where everybody knew his place and obeyed blindly instead of using his own capabilities. The 'identity' movements of today particularly emphasize the limits of individualism in the name of 'belonging to a cultural tradition', or even a race. Rightist 'White supremacists' and present-day Islamists advocate that western individualism is in fact decay and chaos, a claim that was recently voiced as well in Russia's official appreciation of the 'democratic ways in the West'.

Instead of siding with one or the other 'moralism' in this case, I take the stand that a deeper problem with the conceptualization of morals and morality might be the issue here. In my contacts with fellow humanists and Enlightenment defenders, I understood the emphasis in humanistic public statements is quite often exclusively on the individual's right to decide for herself, unhampered by others. Thus, endless discussions on the removal of the headscarf for Muslim women, or on freedom of speech (including the right to offend others, as in some 'woke' circles at least) were engaged in because of this exclusive focus on the individual. I understand rebellion and freeing oneself of external chains (i.e. 'negative' freedom), but I feel more and more uneasy about the exclusivity of that attitude. The mere recognition of the need to cut oneself loose cannot possibly suffice as a solution in moral issues. The main element lacking here (somewhat recognized in the question about 'positive' freedom, in the old distinction) is that of responsibility. The well-documented history of the freedom notions in de Dijn (2020), as well as the overview of contemporary attempts to fill in responsibility notions in the complex, interdependent world of today (in Nelson 2022), point to this blind spot in 'traditional' humanism, I would say.

The historical analysis, tracing freedom all the way back to Ancient Greece, has already made clear that the pendulum today has been swinging away from any restraints by a (larger) group on individual prerogatives. This means that responsibility for others, for the future generations and for 'nature' is considered to be dependent on the free will of the individual: 'if I do not see the need to consider any of the above in my choices, then it is my right as an individual to do so'. Again, during the pandemic we were living in for almost three years and in the face of absolutely devastating climate developments, that is the position taken by deniers, anti-vaxers, believers in conspiracy theories and suchlike – and indeed by several humanists. The claim most often heard is that 'nobody has the right to attack my freedom'. De Dijn (2020) reminds us that the old view in Athens recognized responsibility in a particular way: after free discussions on possible choices a majority vote would decide what 'Athens' would be doing, and each and every individual at the gathering was from then on held to defend that compromise regardless of his (only men at the time!) individual preference. This aspect is absent from the contemporary voices (like the anti-vaxers and others). Nelson (2022) adds a new dimension by pinpointing different, but largely similar, avenues of thought that steer forces, groups and environmental movements when claiming 'freedom'. Her analysis is intriguing since she traces such open-minded reactions in more or less marginal, or at least not mainstream, groups (queer people, drug users and suchlike).

In my opinion, the need to expand the moral frame beyond the individual is imminent: the crises we have triggered as 'free acting economic players' (under the banner of the so-called free market) are so overwhelming that we need to consider our responsibility vis-à-vis nature and others unavoidably and urgently. Individualism in itself does not suffice. Surprisingly, on the basis of biological sciences this expansion of responsibility to nature and other human traditions can draw on the systems approach emanating from such forerunners as Ludwig von Bertalanffy, and Francisco Varela and Humberto Maturana. Presently, this approach can be found in such social-cultural movements as the 'integral life' initiative (Robb Smith: https://vimeo.com/693897790/23789beb01) and in the influential work of Jeremy Lent (2017). These authors and the (many) organizations that are circling around them seek to combine a respectful, inclusive global perspective and responsibility vis-à-vis nature with new suggestions for humanistic morality. In this trend in analysis, when it comes to making moral choices for the survival of all of humanity, the cautious and indeed often holistic perspective of so-called 'others' is considered to be of equal status to the western ones (both those of the religions of the book and those of humanism and Enlightenment).

On top of that, nature is held to show regularity, self-regulating mechanisms and a certain amount of self-organization (see the theories of Varela and Maturana of the 1980s, but also that of the Santa Fe Group in Kauffman 2010). These aspects of nature make it something different from the blunt 'inert nature' of the religions of the book and of the Eurocentric outlook

Reset or the Extinction of *Homo Sapiens?* • 157

up to this day. This should allow for an elaboration or reset of the humanistic perspective and invite us to 'reserve a place at the table for nature' rather than treat it as inert and at the mercy of the only willing species one could think of, i.e. humanity. When sharing a place at the table with other cultures and with nature, we will be more inclined (I hope and think) to listen to them and start reasoning on the responsibility necessary for the whole, consisting of humans of all kinds and nature in one interconnected 'horizontal' network of dependencies and opportunities. The factual interdependence that is already our reality today will become the starting point of a morality that takes interdependence as a basic and indeed founding value for the survival of all. It is obvious that, from such a perspective, the mere and unconditional appeal to freedom of one partner in the whole (i.e. humans of western origin) will not do.

Let me give a concrete example: greed is an awkward and strangely 'counter-natural' attitude in such a view. Hence, instead of being hailed as a right and basic value (as some neoliberals do), it will be considered as at best a mistake and at worst a crime in this new view. Or, another instance: raising children to be first and foremost the best in the so-called competitive society of globalization is promoted as an ideal and presented as a rule for good pedagogy today (e.g. by the OECD). In the alternative world of interconnectedness it will become a serious hindrance to the growth of adult and morally developed humans: victory over the 'other' or over nature will be a chimera because, in the long run, this emphasis on what can be seen as narcissism may primarily harm the other partners in the network and thus become a threat to the survival and possible good life of all. Of course, the field to be discussed here is much larger: these two examples just problematize what seem to be more or less common choices today.

These remarks lead me directly to the political level.

C. Political Changes

When I state that the group and the community matter, and even more that the community should be widened to include non-human partners (formerly seen as 'inert nature'), it is clear that I move up to the political level. I am conscious, of course, that humanism and politics are not often discussed in the same train of thought, since humanists (in Europe and elsewhere, as far as I know) shy away from political choices to a large extent. I have heard they do so because political choices are often divisive – e.g. between social democrats, socialists, anarchists, free-market believers and Christian democrats – while all of them can see themselves as humanists. However, in the most noble and genuine sense, 'politics' refers to choices, values and insights at the level of communities instead of (only) individuals, and that dimension is increasingly unavoidable in any attempt to update the philosophical stand on humans, life and the world that we call humanism. The lack of at the very least minimal limits to individual choice has us land in a world situation which promises hardship and quite terrible living conditions for millions and a deep ecological crisis for many species. So, political rules and contracts seem unavoidable, although they will certainly limit the prerogatives of any individual. It is only within these limits that individuals (and groups and communities) can choose freely.

What are these 'universal' limits? The human rights project can be seen as a forerunner, I suggest. It tried to delineate limits on how humans can be treated and what rights all human beings should have regardless of ethnic, cultural, religious or any other particular makeup they exhibit. Something along these lines should be agreed upon by all human communities for non-human natural phenomena as well. And it should be binding, so that it can effectively be enforced on any and all human actors and their products (I am thinking of ICT developments here, in the light of the singularity prognosis). Of course, specialists in legal matters will comment that even human rights are not enforceable in a sufficient, let alone a general, way. But that means only that work has to be done here – and with the high probability of generalized human suffering because of humans' rash ways of treating each other and nature, the sense of urgency to engage in this work should be obvious.

I do not go into the recent discussions by philosophers on political thinking and action as such. Leftist critiques draw on Karl Marx, Rosa Luxemburg or moderate socialists, while the conservatives feel aligned with Ludwig von Mises or Friedrich von Hayek. Finally, I will not sketch the recent school of thought, mainly in France, which explicitly refers to the Paris Commune of 1871, with such groups as 'Comité invisible' or Mabo (Alain Badiou, Giorgio Agamben, and others, drawing on Gilles Deleuze and Félix Guattari's ideas to some extent: Lesage 2022). This all makes for very interesting philosophizing at a meta-level, but so far it is nothing more than that: in an astonishing way – at least for me – such thinkers do not study any concrete community initiatives like commons and cooperatives, but engage primarily in a meta-discussion on theories and principles of modernity, society or economics – all of it based on texts and textual critiques. The feeling one gets is that Edward Said's remarks on the textuality bias have not been digested yet. For me, this might be an interesting subgroup of intellectual action that needs studying, but it is unlikely to have a deep appeal for large populations. Therefore, I will not go into it here.

The obvious question will then be: but we tried with universal rights and it did not work out because the world has never been as endangered as it is now, right? Some authors have already started speculating about the 'extinction' of humanity, or about the organization of 'a great dying' (of African populations or Asians, in a cynical way) in order to 'correct the balance' without having to change ideological values or the consumerism that has been rampant in the West.

In the light of such cynical invitations (see, e.g., Ghosh, 2022) my argument in this book is at least a double one:

1. Stop saying to all individuals and communities who differ from 'us' (westerners, European-stock Christians and humanists) how they should work to become all the same – i.e. become like 'us'. This was ill-based imperialism and produced more inequality and, certainly nowadays, extreme and damaging selfishness. With that: stop building ever-more hierarchical, top-down institutions with their huge outgrowth of legalistic sophistication, working as they have for the benefit of the happy few over and over again in history. This may turn out to be the worst possible road for many, maybe even for all, if you stick to it long enough. On top of that, the universal, hierarchical organization of the world according to the one superior insight (ours, to be sure) is totalitarian in itself. It suffices to agree on a small set of values and norms (see below), and globalization did not automatically lead to a better life.

   Start listening to what the depreciated traditions have been saying about living with each other and with nature, instead of only 'extracting' everything from them and their environment. Surprisingly enough, this will result in more caution instead of pushing for global competition, and maybe more quality of life and less awe for quantity (in the line of 'Less is more', maybe: Hickel 2021).

2. Start looking at realities basically as an environment of circles, embedded in encompassing circles. The doughnut model (Raworth 2017) on economy comes to mind. Think of reality that way, like systems theories do: there are all sorts of local and more general equilibrium systems in reality (Lent 2017). When you start pushing, changing, maybe using (consuming, etc.) one part you change its equilibrium, and in doing so will impact on other parts of reality as well. So 'using' should be careful and responsible, with rights and opportunities for all phenomena involved in the system concerned. Careful rather than damaging use – through modest and cautious growth sometimes, through immediate and continuous safeguarding of the system at several levels most of the time – should be a central principle for human involvement. Agreeing on such a deep principle will certainly limit free-market private-ownership realms. The circular economic projects come to mind here as possible avenues for economic action. At the local level this means that

circular economies, like commons, should not only be allowed but also protected against predatory economic actions and structures at a higher level of extension (like present-day multinationals). One necessary and absolute rule might be that private ownership (of patents, land, money, labour) should be very drastically limited for any individual and human agent at any level (group, organization, country). Serious negotiations on the span of the so-called free market and private ownership (Piketty 2019) should involve peoples from diverse cultural backgrounds. Such values may prove outdated, leaving responsibility for the longer-term consequences undiscussed. I propose this could become a follow-up step in the agelong trend of 'inventing humanity' (Stuurman 2010): such worldwide initiatives of listening and searching for a minimal common set of values.

My inspiration is that, basically, human experiences in all their variety are all we have got from which to start reworking the survival perspectives of the species. The invitation is to start, this time, from culturally inclusive and interconnected (humans + nature) premises. To give a concrete example, courageous lawyers have been sewing governments and multinationals for the violation of the principle 'duty of care'. The best-known legal fighter is probably Roger Cox: he filed cases against the Dutch government, and won. His example was followed by German, American and other law suits by fellow lawyers–activists. His book (Cox 2020) calls on people to start a 'legalistic revolution': hold governments and multinational corporations legally responsible for their neglect of the duty of care, when inadequate rulings about climate problems progressively attack the wellbeing of coming generations. His work was publicized forcefully in the news media, and he was recognized by former Vice-President Al Gore as an exemplary climate activist.

Beyond this particular legal focus, we can start using knowledge about different traditions to start an open-minded comparative study and negotiation on possible global values and norms. The basic question will be: can humanity alter its disastrous course? With the historian Siep Stuurman (2010), we can appreciate the past three thousand years as the 'invention of humanity', in the sense that ever-wider and more diverging groups and traditions were gradually included in what is now the perspective of the one human species. The next step will be to have all understand that we have become interdependent in an irreversible way. Over the past three thousand years the conceptualization of humanity grew through bloody conflicts, yielding only recently an institution recognizing equal rights for all humans. The result proves insufficient. Let's work on it and take the trio of global interdependence, human diversity and interconnectedness as factual starting points.

If we look at the earthly reality as conscientious humanists of today, we can thus start thinking from a perspective of earth as an ever so much intertwined or interdependent, more or less harmonious, stable system of equilibrated circles. Economically, we are interdependent (certainly in the urbanized world, holding more than 60 per cent of the world's population now), but most of all this factual interdependence became visible as an intrinsic characteristic of ecological and climatic human–earth relationships through the crises referred to. Some strong and binding rules should be negotiated about what the absolute limits of use and other types of interactions should be, never to be transgressed. Again, not only all of humanity – in its cultural diversity – should be respectfully and fully participating in these negotiations, but also natural or non-human phenomena. The rights of oceans, water, air, natural resources, animals, plants and so on should be guaranteed and defended in this scope on interdependence for survival.

Within these limits local, circular choices in the survival view will most probably flourish: stop saying that we should all eat beef or rice all of the time, all over the planet, provided by the same multinational food chain. The commons model may well become the first type of economy for survival, instead of the relatively marginal phenomenon it is now. The reason is obvious: certain resources grow and flourish in certain soils in particular climates, and so on, and the local equilibrium should be the premise here rather than the international profits of the cheapest or the 'cleverest' provider. Multinational agents may possibly have a role for certain products or services, but their action radius will be limited by general agreement. When people object that such is an authoritarian system of economic policy, the counterargument reads: when a global pandemic struck the world (including the West), almost automatically the defenders of free-market ideology withdrew from the forefront and had states define what essential services and essential survival needs and forms prevailed, even in the democratic countries. Almost nobody objected at that point to a serious reduction in the space and power of major market players: many of the presumed important businesses allowed governments to almost unilaterally organize survival. The latter opted for sustaining local, often circular, services like healthcare, education, food supply and so on, and not for luxury goods or any other 'free choice' privileges. In a sense, this was a massive experiment in societal organization, where it appeared that those countries where privatization of healthcare and education were a priority, for example, and which depended more on so-called global free-market supplies (I think of the struggle over Obamacare in the USA) had proportionally less success in survival for all than countries with more limited power of the market players. This 'experiment' should be analysed from the perspective of the global and deeply threatening crises we have landed in now and for coming generations. My plea is that humanists of today should consider working out a clear and all-encompassing scope on how and to what extent such global limits to individual choice can be drawn from the occurrence of such crises: responsibility for the welfare of humans and of natural phenomena should define important limits for individual freedom. The famous rule that 'man is the measure of things' will have to be rephrased accordingly.

These are just some very impressionistic hints at what could, and I think what should, be done by the contemporary humanist. I hope my fellow humanists agree at least on the urgency, and start developing the sort of negotiation groups and the inclusive formats that I have hinted at.

Conclusion
The End, for Now

Of course, this is a critical book. It is mildly optimistic also. If we start renewing a self-critical analysis of one of the great traditions of the world, i.e. the humanism–Enlightenment movement emanating from Europe's fifteenth- to sixteenth-century culture, we will make a positive contribution to the adventure we are drawn into. That adventure will most likely have us experience reality on a continuum between two extremes: at one extreme there is the downfall and harsh conflict period that we are blindly causing by what is euphemistically called globalization, according to one perspective on humanity. The other extreme could be described as the conscious redressing of the cautious, inclusive and interconnected human–nature balanced system of interdependence and impact.

Indifference and denial seem to be all around in the rich countries and in the elites of poorer parts of the world for most of the time: researchers who warn that we are causing worldwide disaster by the way we live, with the privilege of so-called freedom of individual choice in the so-called free market, are still framed as marginal or anti-progress prophets. Eco-realists, believers in conspiracy theories, neoliberals and religious fundamentalists seem to have more influence so far in their attempt to hinder self-criticism and the transition towards a more durable and more just relationship with others and with the earth as a whole. But on the other hand, small-scale commons are growing all over the globe, and interest in alternative views to the disastrous extractivism of the West is expanding. So, there is hope that we can at least avoid total self-destruction. I see a reset of humanism as discussed here as a useful and indeed necessary contribution of the West (those with the primarily European legacy) to the pressing global debate that is emerging, provided we start working on this critical and honest self-reflection now. This will be relatively painful for some: we will have to learn to listen (especially to the former 'primitives'), to become genuinely inclusive, to build down the privileges of the individual and agree on a liminal and sufficient set of values at the level of the earth and *Homo sapiens* as an interdependent species.

In that sense, the end of this book can only be a next stage in an ongoing process of change and readjustment. And time is of the essence.

Afterword

Tim Ingold

These are difficult times for humanism. What had once been hailed as a driving force for the progress and emancipation of peoples from around the world has come under sustained attack from both the right and the left of the political spectrum. From the right, humanism's defence of individual freedom, liberal-democratic governance and the universal power of reason is routinely dismissed as the self-serving posturing of a cosmopolitan and meritocratic elite, determined to corner the benefits of wealth creation for itself while riding roughshod over deeply held feelings of attachment to faith, ancestry, community and nation among those left behind in the race to the top. The shameless exploitation of the grievances of the left-behind, above all by politicians concerned to protect their own ill-begotten and often inherited wealth and privilege, largely accounts for the rise of right-wing populisms in many countries around the world. Meanwhile from the left, humanism has been exposed as having provided ideological cover for a world-historical project of colonialism that has brought untold affluence for some – primarily in Europe, North America and Australasia – at the expense of the impoverishment, enslavement or even genocide of others, the appropriation their lands and the devastation of their environments. Newer pathogens of humanism, such as corporate neoliberalism and algorithmic surveillance, are only exacerbating the situation, and we are already seeing the results in the form of climatic disruption, pandemic disease and cyberwarfare. For its left-wing critics, indeed, humanism is an ideology so infected by its association with colonialism that no postcolonial future is conceivable is not posthumanist as well.

Opponents of humanism, however, should perhaps be careful what they wish for! Do we really want a world riven by competing fundamentalisms,

166 • Afterword

each intolerant of variation in its midst, pushing all difference to an absolute boundary between 'us' and 'them', and placing a patriotic duty on every inhabitant to protect the soil of the nation and its denizens from alien incursion? Such a barricaded world, as we know not only from history but also from today's armed conflicts, is hardly conducive to human flourishing. But at the other extreme, would we really want to discard the idea of universal human betterment? It is all too easy for intellectuals of a posthumanist persuasion – who come predominantly from affluent countries and have enjoyed all the benefits of literacy, education, improved healthcare and democratic governance – to pronounce that the days of humanism are now over. Would they, then, prefer a world without libraries, without colleges and universities, ravaged by preventable disease and ruled by populist demagogues? This is no fantasy nightmare, since we are already seeing these things coming to pass before our very eyes, and seem to lack the mettle to prevent them. And what are we to say to the millions of people around the world who are still denied access to even basic education, not to mention medical care and democratic representation. Are we to tell them that the ship of humanism has already sailed; that we are in a posthumanist world now? Are they to be left stranded, washed up by the tide of history?

That's why the manifold crises of our age demand a fundamental rethink of humanism, not a rejection of it. As Pinxten teaches us, we are in urgent need of a *new* humanism, the newness of which lies less in a repudiation of the past than in its rejuvenation. Part of the problem with the old humanism, indeed, lay precisely in its thirst for innovation, its insistence that the only way forward is to turn one's back on what has gone before. As generation after generation has claimed the present for itself, seeking in the name of progress to outdo the designs, inventions and discoveries of its predecessors, it has left an accumulating pile of ruination in its wake. The latest conceit is to suppose that we are at the dawn of a new epoch, not just in the career of global humanity but in that of the very planet it has now fully colonized. For better or worse, the fate of the earth and all its creatures is seen to lie in human hands. For prophets of the Anthropocene, it is not just previous generations that we have to put behind us, but the entirety of human history. Having fulfilled their historical destiny, they say, humans are poised to become masters of the earth only perhaps to find that they have released forces, of a magnitude beyond their control, which threaten oblivion. This is only the latest version, however, of an old story. Theorists of progress have always been inclined to believe that their own generation is on the point of opening the last envelope.

What, then, is the alternative? Where, Pinxten wonders, did the old humanism go wrong? As we reach the end of this book, it seems pretty clear that its fate was sealed from the moment when philosophers of the Enlightenment, in the Europe of the seventeenth century, determined that the future for humankind lay not in following the footsteps of ancestors or in aligning the ways along which human lives are led with the ways of the animals and plants with which they coexisted, but in turning around to face in the contrary direction. With that, both human ancestral traditions and the lifeways of nonhuman species were consigned to the past, as a heritage to be conserved rather than as a fount of renewal. Nature lost its potential of natality, of giving birth to a world; culture lost its potential of cultivation, of creating the conditions for newborn nature to flourish. If this analysis is correct, then the alternative is clear. It is to turn around once again, to cast our lot in with the lives of those who have gone before us, and those living around us, facing the same direction as they, in order to find a path together into the future. It is, in short, to rejoin the very ways on which we had once turned our backs. This means relearning the arts of coexistence, not just with our fellow human beings but with other creatures as well. And it means exchanging the time of chronological succession, as generation replaces generation, with the duration of life itself, as it continually begets new forms.

Such would be a new humanism. Yet how would it respond to the charges of anthropocentrism and Eurocentrism so often levelled against the old? On closer inspection, neither charge really stacks up. On the first, thinkers of the Enlightenment, far from placing Anthropos at the centre of the cosmos, set out to achieve the very opposite – that is, to *decentre* the human, above all by detaching the faculty of reason, wherein the essence of humanity was supposed to reside, from any worldly involvement whatsoever and placing it instead at the apex of a pyramid with the world at its feet. From this sovereign viewpoint, they supposed, the entirety of nature was revealed to human oversight. And on the second charge, the appeal to the universality of reason, which lay at the heart of the Enlightenment project, repudiates any affiliation to place, region or continent. Reason, being everywhere, belongs nowhere. If everyone is equally endowed with reason, as humanists believe, then it should in principle make no difference where they come from. True, most humanists of earlier centuries assumed that reason was cultivated to different degrees among different nations, on a scale from primitive to civilized, and that in the nations of Europe from which they themselves predominantly came it was cultivated to the highest degree. That they were biased towards their own kind is undeniable. But this bias was not an intrinsic part of their philosophy, as it was, for example, for evolutionary materialists who, following Darwin, had linked degrees of ascendancy of reason over instinct to innate variations in intellectual capacity.

But if the old humanism was literally innocent of both anthropocentrism and regional attachment, for the new humanism these would prove foundational. It would acknowledge that humans are indeed exceptional among animals in their gift of speech, which uniquely grants them the capacity not just to tell the stories of their lives in the living of them but also to interweave their own life stories with those of human and nonhuman others into a Story for the world. This Story is, in effect, an ever-unfolding conversation, and the special responsibility of humans is to open it respectfully to beings of every possible kind. Undivided by barriers of inclusion and exclusion, the conversation is universal, yet its universality – its oneness, if you will – is quite unlike the universality of reason. For in the language of the old humanism, reason speaks with one voice, and one alone, regardless of the identity or background of the speaker. The conversation, however, encompasses as many voices as it has participants, and each is different, continually brought forth in its solicitation of the voices of others. Every voice, positioned somewhere in the world, resounds to the depth of the region to which it belongs, and of the relations it entertains with co-inhabitants. In this sense, the conversation is not so much universal as pluriversal, one-in-many and many-in-one. It would be the vocation of a new humanism to listen to these voices, and to heed their wisdom. This can only be done, however, by taking up a position at the very centre of a worlding world, in the midst of everything. Could the name of this new humanism, indeed, be *anthropology*?

Tim Ingold is Professor Emeritus of Social Anthropology at the University of Aberdeen. His more recent work explores environmental perception and skilled practice. Ingold's current interests lie on the interface between anthropology, archaeology, art and architecture. He is a Fellow of the British Academy and the Royal Society of Edinburgh. In 2022 he was made a CBE (Commander of the Order of the British Empire) for services to anthropology.

References

Albright, Madeleine. 2018. *Fascism: A Warning*. New York: Harper Collins.
Arkoun, Mohammet. 1982. *L'humanisme arabe au IV-Xème siècle (Arab humanism in the IVth-Xth century)*. Paris: Vrin.
Atweh, Ben, Mary Graven, Will Secaba and Paola Valero (eds). 2011. *Mapping Equity and Quality in Mathematics Education*. New York: Springer.
Baker, Michael. 2023. 'The Western Mathematic and the Ontological Turn: Ethnomathematics and Cosmotechnics for the Pluriverse', in Eric Vandendriessche and Rik Pinxten (eds), *Indigenous Knowledge and Ethnomathematics*. New York: Springer, pp. 243–76.
Bauwens, Michel. 2013. *De wereld redden (To save the world)*. Antwerpen: Houtekiet.
Ben Chikha, Chokri. 2017. *Zoo Humain*. Tielt: Lannoo.
Benveniste, Emile. 1969. *Le vocabulaire des institutions indo-européennes (The Vocabulary of the Indo-European Institutions)*. Paris: Edition de Minuit.
Bernabé, Jean. 1981. *Le symbolisme de la mort (The symbolism of death)*. Ghent: CC publications.
Bodelier, Ralf. 2021. *Kosmopolieten (Cosmopolitanists)*. Turnhout: Gompel & Svacina.
Borneman, John. 2019. 'Opposition and Group Formation: Authoritarianism Yesterday and Today', in Joanna Cook, Nicolas Long and Henrietta Moore (eds), *The State We're In: Reflecting on Democracy's Troubles*. Oxford: Berghahn, pp. 97–121.
Bourdieu, Pierre. 2004. *La misère du monde*. Paris: Edition de Minuit.
Boyd, Doug. 1974. *Rolling Thunder*. New York: Delta Books.
Callebaut, Werner and Rik Pinxten (eds). 1987. *Evolutionary Epistemology: A Multiparadigm Program*. Dordrecht: Reidel.
Campbell, Donald T. 1989. *Methodology and Epistemology for Social Sciences*. Chicago: Chicago University Press.
Chronaki, Anna. 2011. 'Disrupting "Development" as the Quality/Equity Discourse: Cyborgs and Subalterns in School Technoscience', in Ben Atweh, Mary Graven, Will Secaba and Paola Valero (eds), *Mapping Equity and Quality in Mathematics Education*. New York: Springer, pp. 3–14.
Chronaki, Anna and Eirini Lazaridou. 2023. 'Subverting Epistemicide through "the Commons": Mathematics as Re/Making Space and Time for Learning', in Eric Vandendriessche and Rik Pinxten (eds), *Indigenous Knowledge and Ethnomathematics*. New York: Springer, pp. 161–82.
Cole, Michael. 1996. *Cultural Psychology*. Cambridge, MA: Harvard University Press.
Commers, Ronald. 1992. *Het vrije denken (Free Thinking)*. Brussels: ASP.

170 • References

———. 2009. *Kritiek van het ethisch bewustzijn. Van liefde met recht en rede (Critique of ethical consciousness: From love with rights and reason)*. 2 volumes. Louvain: Acco.

Condoni, Sylvie and François Savatier. 2019. *Néandertal, mon frère: 300 000 ans d'histoire (Neandertal, my brother: 300,000 years of history)*. Paris: Flamarion.

Cook, Joanna, Nicholas Long and Henrietta Moore (eds). 2019a. *The State We're In: Reflecting on Democracy's Troubles*. Oxford: Berghahn.

———. 2019b. 'Introduction: When Democracy 'Goes Wrong', in Joanna Cook, Nicolas Long and Henrietta Moore (eds), *The State We're In: Reflecting on Democracy's Troubles*. Oxford: Berghahn, pp. 1–26.

Corijn, Eric. 2017. *Kan de stad de wereld redden? (Can the city save the world?)*. Brussels: VUB Press.

Cox, Roger. 2020. *Revolution Justified*. Maastricht: Planet Prosperity Foundation.

Dalsheim, Joyce and Gregory Starrett. 2021. 'Everything Possible and Nothing True: Notes on the Capitol Insurrection', *Anthropology Today* 37(2): 26–30.

Davis, Michael. 2006. *Planet of Slums*. London: Verso.

de Dijn, Annelien. 2020. *Freedom: An Unruly History*. Cambridge, MA: Harvard University Press.

De Ketelaere, Georgette. 2020. *Mens versus Machine (Humans against machines)*. Antwerp: Pelckmans.

de la Cadena, Marisol. 2011. *Earth Beings: Ecologies of Practice across Andean Worlds*. Durham, NC: Duke University Press.

De Munter, Koenraad. 2007. *Nayra: Ojos al sur del presente*. Oruro, Bolivia: CEPA.

———. 2010. *De culturele eeuw (The cultural century)*. Antwerp: Houtekiet.

Descola, Philippe. 2005. *Par-delà nature et culture (Beyond nature and culture)*. Paris: Folio.

———. 2016. 'Biolatry: A Surrender of Understanding (Response to Ingold's "A Naturalist Abroad in the Museum of Ontology")', *Anthropological Forum* 26(3): 1–8.

———. 2021. *Les Formes du Visible (The Forms of the Visible)*. Paris: Editions du Seuil.

Dobbelaere-Welvaert, Mathias. 2020. *Ik weet wie je bent en wat je doet (I Know Who You Are and What You Do)*. Ghent: Borgerhoff & Lamberigts.

Dumolyn, Jan and Arthur Brown. 2019. *Medieval Bruges: 850–1550*. Cambridge: Cambridge University Press.

Eggers, Dave. 2021. *Every*. New York: Vintage.

Einstein, Albert. 1949. *Autobiography*, in Paul Schilpp (ed.) (1969), *Albert Einstein: Philosopher-Scientist*. Lasalle, IL: Open Court, pp. 1–96.

Escobar, Arturo. 2017. *Designs for the Pluriverse: Radical Interdependence, Autonomy, and the Making of Worlds*. Durham, NC: Duke University Press.

Faassen, Vesna and Lukas Verdijk. 2017. *Wanneer we spreken over kolonisatie (When we talk about colonization)*. Brussels: Public Actions.

Fabian, Johannes. 1979. 'Interview', in Rik Pinxten (ed.), *On Going Beyond Kinship, Sex and the Tribe*. Ghent: Story, pp. 56–83.

———. 1984. *Time and the Other*. New York: Columbia University Press.

Farella, John. 1984. *The Main Stalk: A Synthesis of Navajo Philosophy*. Flagstaff Ariz.: Arizona University Press.

Fassin, Didier. 2021. 'The Jungle of Calais', *Anthropology Today* 37: 1–4.

Fawcett, Edmund. 2015. *Liberalism: The Life of an Idea*. Princeton, NJ: Princeton University Press.

Fisher, Laura. 2017. 'Why Shall We Have Peace to Be Made Slaves?', *Ethnohistory* 64: 91–114.

Fitzgerald, Timothy. 2021. 'A Critical View of Anthropology of Religion', *Anthropology Today* 37(2): 1–3.

Frankopan, Peter. 2018. *The New Silk Roads*. London: Bloomsbury.

————. 2023. *The Earth Transformed: An Untold History*. London: Bloomsbury.

Ghosh, Amitav. 2022. *The Nutmeg's Curse: Parables for a Planet in Crisis*. Chicago: John Murray.

Gilbreath, Karl. 1973. *Red Capitali$m: An Analysis of the Navajo Economy*. Norman, OK: University of Oklahoma Press.

Goodman, Nelson. 1978. *Ways of Worldmaking*. Boston, MA: Hackett.

Goorden, Lea. 2019. *De essentie van Arendt (The Essence of Arendt)*. Leusden: ISVW.

Gorski, Philip S. 2021. 'Right-Wing Populism and Religious Conservatism: What's the Connection?', in Florian Höhne and Torsten Meireis (eds), *Religion and Neo-Nationalism in Europe*. Baden Baden: Nomos, pp. 333–46.

Graeber, David. 2013. *The Democracy Project*. London: Penguin.

————. 2015. *Bullshit Jobs*. London: Penguin.

Graeber, David and David Wengrow. 2021. *The Dawn of Everything*. London: Penguin.

Harari, Yuval. 2015. *From Animals into Gods: A Brief History of Humankind*. Or Yehuda, Israel: Kinneret Zmora-Bitan Dvir.

Hari, Sandew. 2023. *Decolonizing the Mind: A Guide to Decolonial Theory and Practice*. The Hague: Amrit.

Hickel, John. 2021. *Less Is More: How Degrowth Will Save the World*. London: Penguin, Random House.

Hitchcock, Robert K. and Kathleen A. Galvin. 2023. 'Hunter-Gatherer Societies: Ecological Impact of.' In: *Reference Module in Life Science*. Amsterdam: Elsevier.

Höhne, Florian and Torsten Meireis (eds). 2021. *Religion and Neo-Nationalism in Europe*. Baden Baden: Nomos.

Huntington, Samuel. 1996. *The Clash of Civilizations*. New York: Basic Books.

Hymes, Dell. 1978. *American Structuralism*. Philadelphia, PA: Pennsylvania University Press.

————. 1981. *In Vain I Tried To Tell You*. Philadelphia, PA: University of Pennsylvania Press.

Ingold, Tim. 2004. *The Perception of Environment*. London: Routledge.

————. 2016. 'A Naturalist Abroad in the Museum of Ontology: Phillipe Descola's *Beyond Nature and Culture*', *Anthropological Forum* 26(3): 301–20.

————. 2017. 'The Unsustainability of Everything'. www.che.ac.uk.

Kauffman, Stuart. 2010. *The Reinvention of the Sacred*. New York: Basic Books.

Kissinger, Henry. 1994. *Diplomacy*. New York: Basic Books.

Kruithof, Jaap. 2001. *Het humanisme (Humanism)*. Antwerp: EPO.

Kuhn, Thomas. 1962. *The Structure of Scientific Revolutions*. Chicago: University of Chicago Press.

Kurzweil, Raymond. 2020. 'The Coming Singularity | Big Think'. Video. YouTube. Retrieved 1 September 2023 from https://www.youtube.com/watch?v=1uIzS1uCOcE.

Lave, Jane and Sylvia Scribner. 1986. *Situated Learning*. Cambridge: Cambridge University Press.

Lent, Jeremy. 2017. *The Patterning Instinct*. New York: Prometheus Press.

Lesage, Dieter. 2022. *Het parlement en de velen: Pleidooi voor radicale democratie (The Parliament and the Many: Plea for Radical Democracy)*. Antwerp: Houtekiet.

Lévi-Strauss, Claude. 1958. *Anthropologie structurale (Structural Anthropology)*. Paris: Editions du Seuil.

————. 2016. *De Montaigne à Montaigne (From Montaigne to Montaigne)*. Paris: EHESS.

Levitsky, Steven and David Ziblatt. 2018. *How Democracies Die*. New York: Viking.

Long, Colin. 1964. *Alpha*. New York: Doubleday.

Lotens, Walter. 2019. *Rebelse plekken: Over municipalisme en commons (Rebellious places: On municipalism and commons)*. Turnhout: Gompel & Svacina.

172 • References

Lucretius. 2008. *De natuur der dingen (The Nature of Things)*. Groningen: Historische Uitgeverij.

Maeckelbergh, Margaret. 2019. 'Politics after Democracy: Experiments in Horizontality', in Joanna Cook, Nicolas Long and Henrietta Moore (eds), *The State We're In: Reflecting on Democracy's Troubles*. Oxford: Berghahn, pp. 190–212.

Mann, Thomas. 1996. *The Magic Mountain*. New York: Vintage International.

Mattei, Umberto and Laura Nader. 2014. *Plunder: When the Rule of Law is Illegal*. Berkeley, Calif.: University of California Press.

Maturana, Humberto and Francisco Varela. 1980. *Autopoesis and Cognition: The Realization of the Living*. Boston, MA: Reidel.

McNeley, James. 1981. *Holy Wind in Navajo Philosophy*. Flagstaff, AZ: Arizona University Press.

Meireis, Torsten. 2021. 'Religious Internationalism? German Protestantism, Neo-Nationalism and Populism', in Florian Höhne and Torsten Meireis (eds), *Religion and Neo-Nationalism in Europe*. Baden Baden: Nomos, pp. 391–406.

Meyer, Herman and Anthony Benavoite (eds). 2013. *PISA, Power and Policy.: The Emergence of Global Educational Governance*. Oxford: Symposium Books.

Mignolo, Walter. 2015. 'Anomie, Resurgences, and De-Noming. Foreword', in Federico Luisetti, John Pickles and Wilson Kaiser (eds), *The Anomie of Earth: Philosophy, Politics and Autonomy in Europe and the Americas*. London: Duke University Press, pp. VII–XVI.

Montaigne, Michel de. 2004. *De essays* (orig. 1595. *Essais*). Amsterdam: Polak & Van Gennep.

Morgan, Lewis Henry. 1851. *League of the Iroquois*. Secausus, NJ: The Citadel Press.

Nader, Laura. 2015. *What the Rest Think of the West*. Oakland, CA: University of California Press.

Needham, Joseph. 1965–2000. *Science and Civilization in China*. Vol. II. Cambridge: Cambridge University Press.

Nelson, Maggie. 2022. *Over vrijheid in kunst, seks, drugs en klimaat (On Freedom in Art, Seks,Drugs and Climate)*. Amsterdam: Atlas Contact.

NIC. 2021. 'Global trends 2040'. Washington DC: Publication of the National Intelligence Councel.

Nussbaum, Martha. 2011. *Creating Capabilities: The Human Development Approach*. Cambridge, MA: Harvard University Press.

Olyslaegers, Jeroen. 2020. *Wildevrouw (Wild woman)*. Amsterdam: Bezige Bij.

Pääbo, Svante. 2014. *Neanderthal Man: In Search of Lost Genomes*. New York: Basic Books.

Pally, Marcia. 2020. 'Why Vote Against Best Interests or Why Is Populism Persuasive?', in Florian Höhne and Torsten Meireis (eds), *Religion and Neo-Nationalism in Europe*. Baden Baden: Nomos, pp. 361–76.

Pauwels, Caroline. 2021. *Ronduit: Overpeinzingen van een possibilist (Squarely: Thoughts of a possibilist)*. Antwerp: Houtekiet.

Perelman, Chaim and Sylvie Olbrechts-Tyteca. 1957. *La nouvelle rhétorique (The New Rhetoric)*. Paris: PUF.

Pfenninger, Karl and Valerie Shubik (eds). 2001. *The Origins of Creativity*. Oxford: Oxford University Press.

Piketty, Thomas. 2014. *Capital in the 21st Century*. Boston, MA: Harvard University Press.

———. 2019. *Capital et Idéologie (Capitol and ideology)*. Paris: Editions du Seuil.

Pinker, Steven. 2021. *Enlightenment Now*. London: Allen Lane.

Pinxten, Rik. 2001. 'La cosmologie navajo et la cosmologie occidentale', *Itinéraires belges aux Amériques. Civilisations* L(1–2): 43–62.

———. 2007. *De strepen van de zebra (The stripes of the zebra)*. Antwerp: Houtekiet.

———. 2010. *The Creation of God*. Zurich: Peter Lang Verlag.

———. 2016. *Multimathemacy. Anthropology and Mathematics Education*. New York: Springer.

———. 2019. *Kuifje wordt volwassen (Tintin comes of age)*. Antwerpen: EPO.

———. 2020. *Humanisme in woelige tijden (Humanism in turbulent times)* Turnhout: Gompel & Scavina.

———. 2021. 'Neo-Nationalism, Religion and Politics of the Right in Belgium', in Florian Höhne and Torsten Meireis (eds), *Religion and Neo-Nationalism in Europe*. Baden Baden: Nomos, pp. 151–62.

Pinxten, Rik, Erik Soberon and Ingrid van Dooren. 1987. *Towards a Navajo Indian Geometry*. Ghent: C&C Publications.

Pinxten, Rik and Karen François. 2011. 'Politics in an Indian Canyon? Some Thoughts on the Implications of Ethnomathematics', *Educational Studies in Mathematics* 10: 1007–37.

Pinxten, Rik, Ingrid van Dooren and Frank Harvey. 1983. *Anthropology of Space: Natural Philosophy and Semantics of the Navajo*. Philadelphia, PA: University of Pennsylvania Press.

Ponsaers, Paul. 2021. *Georganiseerde wanorde (Organized disorder)*. Turnhout: Gompel & Svacina.

Powers, Richard. 2018. *The Overstory*. New York: Norton & Cy.

Prigogine, Ilya. 1969. 'La fin de l'atomisme', *Bulletin de l'Académie Royale de Belgique* 12: 1110–17.

Prigogine, Ilya and Isabelle Stengers. 1984. *Order out of Chaos*. New York: Wiley.

———. 1987. 'The Meaning of Entropy', in Werner Callebaut and Rik Pinxten (eds), *Evolutionary Epistemology: A Multiparadigm Program*. Dordrecht: Reidel, pp. 57–74.

Raworth, Kate. 2017. *Doughnut Economics*. London: Random House.

Relethford, John. 2006. *The Human Species: An Introduction to Biological Anthropology*. Boston, MA: McGraw Hill.

Restivo, Sal. 2021. *Society and the Death of God*. New York: Routledge.

Rubinstein, David. 2004. 'Mathematics in the Pacific', in *ICME Proceedings*, Washington, DC: mimeo.

Sahlins, Marshall. 2008. 'The Corporatization of Academia', *Anthropology Newsletter* 15: 5–9.

Said, Edward. 1978. *Orientalism*. London: Penguin.

Sapolsky, Robert. 2017. *Behave: The Biology of Humans at Our Best and Worst*. London: Vintage Penguin.

Sassen, Saskia. 2016. *Expulsion*. Amsterdam: Contact.

Schilpp, Paul (ed.). 1969. *Albert Einstein: Philosopher-Scientist*. Lasalle, Ill.: Open Court.

Sen, Amartya. 2002. *Rationality and Freedom*. Cambridge, MA: Harvard University Press.

———. 2009. *The Idea of Justice*. London: Allen Lane.

Simpson, Ingrid. 2013. 'Go Offshore, Young Man: Young Millionaire Technologists Challenge Constitutional Culture by Building Autonomous Ocean Communities', *Anthropology News* 54(7): 6–7.

Standaert, Roger. 2021. *The Centralized Test: A Pest?* Brussels: Docville.

Stiglitz, Joseph. 2010. *Free Fall*. New York: Norton & Cy.

———. 2011. 'Of the 1%, by the 1%, for the 1%', *Vanity Fair*, May.

———. 2019. *People, Power and Profits: Progressive Capitalism for an Age of Discontent*. London: Allen Lane.

Stuurman. Siep. 2010. *De uitvinding van de mensheid (The invention of humanity)*. Amsterdam: Bert Bakker.

Susanne, Charles. 2015. *Transhumanisme: A la limite des valeurs humanistes (Transhumanism: at the limit of humanist values)*. Arquennes: Collection Hélios.

174 • References

Thom, René. 1987. 'Epistemology of Evolutionary Theories', in Werner Callebaut and Rik Pinxten (eds), *Evolutionary Epistemology: A Multiparadigm Program*. Dordrecht: Reidel, pp. 97–104.

Tomasello, Michael. 2009. *Why We Cooperate*. Cambridge, MA: MIT Press.

Toulmin, Stephen. 1990. *Cosmopolis*. Cambridge: Cambridge University Press.

van Beurden, Jos. 2017. *Treasures in Trusted Hands*. Amsterdam: Sidestone Press.

Vandendriessche, Eric and Rik Pinxten (eds). 2023. *Indigenous Knowledge and Ethnomathematics*. New York: Springer.

Van Duppen, Dirk and Jan Hoebeke. 2016. *De supersamenwerker (The super-collaborator)*. Antwerp: EPO.

van Yperseele, Jean-Pierre, Theo Libaert and Paul Lamote. 2018. *In the oog van de klimaatstorm (In the eye of the climate storm)*. Antwerp: EPO.

Veyne, Paul. 1989. *Did the Greeks Believe in their Gods?* Cambridge: Cambridge University Press.

Vollmet, Stan Emil, Emily Goren, Chun-Wei Yuan, Jacky Cao and Amanda Smith. 2020. 'Fertility, Mortality, and Population Scenarios for 195 Countries and Territories from 2017 to 2100: A Forecasting Analysis for the Global Burden of Disease Study', *The Lancet* 396: 10258, 1285–1306.

Webster, Anthony. 2015. *Intimate Grammars: An Ethnography of Navajo Poetry*. Tucson, AZ: University of Arizona Press.

Weitz, Eric. 2019. *A World Divided: The Global Struggle for Human Rights in the Age of Nation-States*. Princeton, NJ: Princeton University Press.

Wekker, Gloria. 2018. *White Innocence: Paradoxes of Colonialism and Race*. New York: Duke University Press.

Whitehead, Alfred North. 1919. *An Enquiry Concerning the Principles of Natural Knowledge*. Cambridge: Cambridge University Press.

———. 1929. *Process and Reality*. Cambridge: Cambridge University Press.

———. 1961. *Adventures of Ideas*. New York: Free Press.

Witherspoon, Gary. 1977. *Language and Art in the Navajo Universe*. Ann Arbor, MI: University of Michigan Press.

Wolf, Eric. 1981. *Europe and the Peoples without History*. New Haven, CT: Yale University Press.

Yanagihara, Hannah. 2022. *To Paradise*. New York: Random House.

Zuboff, Shoshana. 2019. *The Age of Surveillance Capitalism*. London: Profile Books.

INDEX OF SUBJECTS

anthropological turn, xviii
biodiversity, xiii, xviii, xviii, 28, 34, 47, 49, 61, 70, 78, 90, 93, 95, 98, 103

capabilities, 12, 51, 138, 148, 150, 153–155, 172
caritas, 20, 21, 79
climate warming, 28, 33, 34, 72, 90, 98, 139, 145, 148
civilization, xi, 5, 16, 31, 40, 59, 60, 67, 114, 116, 135, 146, 171, 172
commons, 26, 27, 34, 74, 76, 78, 158, 160, 161, 163, 169, 171
convivir, 116, 118
cosmopolitanism, viii, 17–18, 24, 84
cultural relativism, xi–xii

decolonization, 22, 47, 62, 81
degrowth, xviii, 78, 126, 171
difference, xxi, 4–8, 12, 85, 113, 166
Diné, 68, 88, 92–95, 104, 114
diversity, 4–8, 42, 55, 79, 85, 100, 113, 150, 154, 160–161
dualism, 12, 42–45, 50, 56, 85, 90–91, 95, 103, 105, 113, 118, 143, 145

ecological, viii, 29, 30, 34, 48, 50, 56, 72, 74, 87–101, 161, 171
ecological destruction, viii, 34
Enlightenment, viii, xi, xii, xvi, xvii, xxi, 8, 11, 17, 18, 22, 36, 51, 54, 56, 61, 64, 65, 67–70, 83, 89, 90, 95, 96, 97, 102, 105, 109, 110, 111, 112, 114, 119, 124, 125, 126, 130, 135, 137, 142, 146, 150, 153, 155, 156, 163, 167, 172
Eurocentric, vii, viii, ix, 56, 57–83, 111, 116, 149, 154, 156
extinction, 5, 53, 106, 118, 127–162
Extinction Rebellion, 26
extractivism, 17, 106, 107, 131, 132, 139, 148, 163

freedom, viii, xx, 8, 12, 17, 19–23, 25, 26, 31, 36, 58–61, 68–72, 77, 79, 80, 85, 91, 96–100, 102, 108–111, 127, 129, 139, 145, 148–150, 155–157, 162, 163, 165, 170, 172, 173
fundamentalism, 22, 165

Generation Identität, 23, 147
Génération Identité, 23
globalization, 24, 25, 31, 50, 56, 61, 66, 77, 119, 144, 157, 159, 163

heteronomic, 138
heteronomy, 15, 19, 125, 133
holism, 49, 50, 76, 85

inequality, xiii, xiv, 6, 20, 24, 25, 33, 34, 40, 47, 61, 63, 67, 70, 80, 97, 107, 108, 111, 132, 137, 142, 150, 159
Indignados, 26

176 • Index

interconnectedness, 46, 48, 77, 80, 94–95, 111, 112, 118, 142, 151, 157, 160
intercultural education, 2, 40
interdependence, xiii–xviii, 6, 24, 28–38, 61, 78–80, 82, 93, 119, 121, 126, 131, 137, 138, 145, 154, 157, 160, 161, 163, 170
IPCC, xiv, 33, 61, 110, 139

liberal democracy, 18–27, 30

Museum of the American Indian, 59–62, 69, 75, 92

Native American, 46, 48, 53, 54, 59–62, 69–70, 79, 80–81, 90, 92, 94, 98, 104, 128, 147, 149, 152, 154
Navajo, 46, 48, 49, 59, 63, 67, 92–94, 114–116, 170, 171, 172, 173, 174
NEED, 39–56, 118, 126, 146, 154
neoliberalism, xi, 6, 12, 17, 25, 34, 66, 67, 106, 107, 128, 165
NIC, 29, 67, 109, 110, 172
nomos, 13, 14, 15

Occupy, 26
Ordo Iuris, 23, 62
ontological turn, xviii, 46, 48, 169

pluriversal, 45, 113, 117, 119, 168
pluriversality, 68, 154
pluriversalism, 80, 117
political economy, 78, 109, 132
primitive, xvi, 14–16, 42, 43, 45, 47, 49, 55, 56, 60, 69, 72, 76, 81, 89, 90, 91, 107, 113, 116, 121, 132, 143, 144, 164
private interest, xiii–xiv, 76, 139
progressophobia, 64, 65, 70

rights, xiv, 12, 17, 18, 22, 31, 32, 43, 44, 58, 70–71, 76, 811, 84, 85, 89, 90, 95, 96–98, 113, 114, 116, 118, 120, 122, 125, 126, 131, 134, 149–150, 155, 158, 159, 160, 161, 170, 174

self-organization, 52, 53, 156
surveillance capitalism, x, xx, 13, 99, 100, 134, 174

tolerance, 12, 21, 59, 68

Umvolkung, 23, 25, 85
universalia, 14, 15, 112, 113
universalism, 12, 13–16, 79, 111, 113, 114, 116, 117, 130–134, 146

woke, 28, 130, 155
worlding, 54–56, 116, 117, 168

Index of Names

Agamben, Giorgio, 158
Albright; Madeleine, 26, 72, 79, 169
Alma, Hans, 11
Apostel, Leo, 11
Arendt, Hannah, 73, 123, 171
Arkoun, Mohammet, 41, 169

Badiou, Alain, 158
Baker, Michael, 153, 169
Bauwens, Michel, 27, 74, 169
Ben Chikha, Chokri, 81, 169
Bentham, Jeremy, 20
Benveniste, Emile, 94, 169
Bernabé, Jean, 87, 88, 169
Biden, Joe, 140
Bodelier, Ralf, 17, 84, 169
Bolyai, Janos, 118
Borneman, John, 27, 169
Bourdieu, Pierre, xviii, 40, 169
Boyd, Doug, 147, 149, 169
Bregman, Rutger, 123
Brown, Arthur, 19, 170
Bruno, Giordano, xv, xvi, 21

Callebaut, Werner, 93, 169, 173, 174
Calvin, John, 97
Campbell, Donald T., 51, 52, 53, 93, 169
Chomsky, Noah, 106
Chronaki, Anna, 62, 169
Clinton, Bill, 106
Cody, Bill, 81
Cole, Michael, 63, 92, 131, 169

Commers, Ronald, 11, 12, 169
Comte, Auguste, 20
Condoni, Sylvie, 5, 170
Cook, Joanna, 13, 26, 169, 170, 172
Corijn, Eric, 27, 170
Cox, Roger, 160, 170

Dagha Chi'lii, 88
Davis, Michael, 39, 170
Dawkins, Richard, 64, 103
De Dijn, Annelien, 22, 31, 96, 97, 124, 155, 156, 170
De la Cadena, Marisol, 46, 49, 75, 93, 94, 95, 170
Deleuze, Gilles, 158
De Munter, Koen, 75, 170
Dennett, Daniel, 64
Descola, Philippe, xix, xxi, xxii, 42, 43, 46, 47, 48, 54, 55, 76, 92, 93, 94, 113, 116, 117, 119, 131, 142, 147, 170, 171
Diderot, Denis, xxi
Dumolyn, Jan, 19, 170

Eggers, Dave, X, 170
Eichmann, Adolf, 73
Einstein, Albert, 150, 170
Erasmus, Desiderius, xv, xxi, 21, 56
Escobar, Arturo, 113, 116, 170

Faassen, Vesna, 61, 170
Fabian, Johannes, xviii, 117, 145, 170

178 • Index

Farella, John, 46, 94, 95, 147, 170
Fassin, Didier, 105, 122, 170
Fawcett, Edmund, 21, 170
Feyerabend, Paul, 124
Fisher, Laura, 16, 60, 79, 91, 132, 154, 170
Fitzgerald, Timothy, 42, 170
Flam, Leopold, 11
Foucault, Michel, 105
Frankopan, Peter, 21, 24, 31, 49, 63, 85, 144, 170
François, Karen, 116, 173

Galvin, Kathleen, 48, 171
Goodman, Nelson, 55, 171
Goorden, Lea, 73, 171
Graeber, David, xviii, xxii, 43, 53, 60, 61, 65, 67, 68, 69, 75, 91, 98, 113, 117, 119, 124, 135, 141, 142, 143, 144, 145, 171
Geertz, Clifford, xviii
Ghosh, Amitav, 15, 106, 107, 128, 131, 143, 144, 159, 171
Gilbreath, Karl, 104, 171
Goldwater, Barry, 104
Guattari, Félix, 158

Harari, Yuval, 141, 171
Hari, Sandew, 77, 117, 171
Harvey, Frank, 46, 94, 173
Heidegger, Martin, 44
Hickel, Jason, xviii, 62, 67, 78, 110, 159, 171
Hitchcock, Robert, 48, 171
Hobbes, Thomas, 16, 68, 69, 70, 125
Hoebeke, Jan, 57, 174
Höhne, Florian, 32, 62, 99, 129, 147, 171, 172, 173
Huntington, Samuel, 4, 40, 71, 82, 171
Hymes, Dell, 94, 171

Ingold, Tim, xviii, 46, 48, 93, 103, 104, 165–168, 171

Kant, Immanuel, xxi, 41, 84, 85, 89, 112, 113, 114, 117, 125, 130, 135
Kauffman, Stuart, 52, 53, 93, 156, 171
Kissinger, Henry, 121, 171

Kruithof, Jaap, 11, 12, 21, 171
Kuhn, Thomas, 65, 124, 171

Lamote, Paul, 33, 61, 103, 110, 174
Lave, Jane, 151, 171
Lazaridou, Erini, 62, 169
Lent, Jeremy, 156, 171
Lesage, Dries, 158, 171
Levinas, Emmanuel, 85
Lévi-Strauss, Claude, xviii, xxi, 92, 171
Levitsky, Steven, 26, 171
Libaert, Theo, 33, 61, 103, 110, 174
Lobatchesky, Nikolai, 118
Locke, John, 6, 97
Long, Nicolas, 26, 90, 169, 170, 171, 172
Lotens, Walter, 34, 171
Lucretius, 103, 172
Luther, Martin, 97
Luxemburg, Rosa, 158

Maeckelberg, Margaret, 27, 172
Mann, Thomas, x, 172
Marx, Karl, 20, 81, 158
Mattei, Ugo, 17, 35, 77, 172
Maturana, Humberto, 156, 172
May, Karl, 60, 81
McNeley, James, 94, 172
Meireis, Torsten, 6, 23, 25, 32, 62, 72, 99, 129, 147, 171, 172, 173
Mignolo, Walter, 14, 172
Montaigne, Michel de, viii, xv, xvi, xxi, 14, 15, 43, 56, 69, 114, 116, 143, 171, 172
Morgan, Lewis Henry, 54, 172
Moore, Henrietta, 26, 169, 170, 172

Nader, Laura, viii–ix, 17, 35, 53, 78, 108, 117, 135, 144, 145, 172
Needham, Joseph, 143, 144, 150, 172
Nelson, Maggie, 155, 156, 172
Nussbaum, Martha, 146, 148, 149, 150, 153, 154, 172

Olbrechts-Tyteca, Sylvie, 45, 84, 114, 124, 136, 172
Olyslaegers, Jeroen, 21, 172
Orban, Victor, 71

Pääbo, Svante, 5, 172
Pally, Marcia, 172
Pauwels, Caroline, 125, 172
Perelman, Chaim, 45, 84, 114, 124, 136, 172
Petrarca; Franscesco, 21, 97
Piketty, Thomas, xiv, 7, 17, 20, 22, 24, 25, 30, 34, 35, 44, 62, 67, 72, 97, 107, 108, 109, 132, 146, 150, 160, 172
Pinker, Steven, 64, 65, 66, 67, 79, 103, 172
Pinxten, Rik, viii, ix, xix, 3, 25, 28, 30, 46, 48, 51, 62, 66, 93, 94, 113, 116, 153, 166, 169, 170, 172, 173, 174
Ponsaers, Paul, 23, 25, 32, 72, 99, 129, 173
Powers, Richard, x, 50, 134, 173
Prigogine, Ilya, 30, 47, 50, 150, 151, 173
Putin, Vladimir, 129, 140

Radcliffe-Brown, Alfred, xviii
Raes, Koen, 11
Raworth, Kate, 27, 34, 67, 78, 118, 159, 173
Rawls, John, 84, 85, 89, 95, 112, 125, 135, 136, 146, 150
Reagan, Ronald, 6, 25, 34, 67
Relethford, John, 5, 30, 173
Restivo, Sal, 55, 173
Riemann, Bernhard, 118
Rousseau, Jean Jacques, 60, 61, 68, 69, 97
Rubinstein, David, 62, 173

Sahlins, Marshal, xvii, xviii, 54, 66, 173
Said, Edward, 83, 85, 158, 173
Sapolsky, Robert, 58, 173
Sassen, Saskia, 24, 173
Schmitt, Carl, 13
Scribner, Sylvia, 152, 171
Sen, Amartya, 112, 125, 135–138, 173
Simpson, Ingrid, 34, 173
Singer, Peter, 44
Smith, Adam, 60, 62, 68, 70
Standaert, Roger, 34, 51, 173
Stengers, isabelle, 30, 47, 151, 173

Stiglitz, Joseph, 7, 17, 18, 25, 30, 32, 67, 72, 108, 109, 173
Stuurman, Siep, 144, 160, 173

Thatcher, Margaret, 6, 25, 34, 67
Thom, René, 151, 174
Tomasello, Michael, 7, 174
Toulmin, Stephen, 84, 114, 124, 125, 136, 174
Trump, Donald, 17, 80, 82, 99, 127, 128

Van Beurden, Jos, 61, 174
Vandenbossche, Marc, 11
Vandendriessche, Eric, 66, 169, 174
Van Duppen, Dirk, 57, 174
Van Dooren, Ingrid, 46, 94, 173
Van Yperseele, Jean-Pierre, 33, 61, 103, 110, 174
Varela, Francesco, 156, 173
Verdijk, Lukas, 61, 170
Veyne, Paul, 174
Vygotsky, Lev, 92
Vollmet, Stan, 36, 174
Von Bertalanffy, Ludwig, 52, 156
Von Hayek, Friedrich, 158
Von Mises, Ludwig, 158

Weitz, Eric, 17, 22, 174
Wekker, Gloria, 61, 174
Wengrow, David, 43, 53, 60, 61, 65, 68, 69, 73, 91, 98, 113, 117, 119, 124, 135, 141, 142, 143, 144, 145, 171
Whitehead, Alfred North, 19, 20, 45, 150, 174
Witherspoon, Gary, 46, 94, 174
Wolf, Eric, 117, 145, 174

Yanagihara, Hannah, 134, 141, 174

Ziblatt, David, 26, 79, 171
Zuboff, Shoshana, 13, 32, 77, 99, 100, 134, 174

www.ingramcontent.com/pod-product-compliance
Ingram Content Group UK Ltd.
Pitfield, Milton Keynes, MK11 3LW, UK
UKHW020719180125
453877UK00007BA/94